OUR ANGRY EARTH

TOR

A TOM DOHERTY
ASSOCIATES BOOK
NEW YORK

OUR ANGRY EARTH

ISAAC ASIMOV
AND
FREDERIK POHL

INTRODUCTION AND AFTERWORD BY
KIM STANLEY ROBINSON

OUR ANGRY EARTH

Copyright © 1991 by Asimov Holdings, LLC and Gateway, LLC

Kim Stanley Robinson's introduction and afterword copyright © 2018 by Kim
Stanley Robinson

A Tor Book
Published by Tom Doherty Associates
175 Fifth Avenue
New York, NY 10010

www.tor-forge.com

Tor® is a registered trademark of Macmillan Publishing Group, LLC.

The Library of Congress has cataloged the first edition as follows:

Asimov, Isaac, 1920–1992.
  Our angry earth / Isaac Asimov & Frederik Pohl.—1st ed.
    p. cm.
  "A Tom Doherty Associates book."
  ISBN 978-0-312-85252-5 (hardcover)
  ISBN 978-1-250-16366-0 (ebook)
1. Environmental protection. 2. Pollution. 3. Nature conservation—
Political aspects. I. Frederik Pohl. II. Title.
  TD170 .A75 1991
  363.7   91019919

ISBN 978-0-7653-9976-2 (trade paperback)

Our books may be purchased in bulk for promotional, educational, or business use.
Please contact your local bookseller or the Macmillan Corporate and Premium
Sales Department at 1-800-221-7945, extension 5442, or by email at Macmillan
SpecialMarkets@macmillan.com.

First Edition: November 1991
First Trade Paperback Edition: March 2018

Printed in the United States of America

0 9 8 7 6 5 4 3 2 1

CONTENTS

INTRODUCTION

Kim Stanley Robinson

Isaac Asimov and Frederik Pohl met in the late thirties in New York, when they were teenagers. Both of them were associated with a group of like-minded young science fiction writers called the Futurians. The Futurians were interested in the Young Communist League and other leftist causes, and this was true to an extent of Asimov and Pohl as well. Both of them later recalled these years very entertainingly in their autobiographies, Asimov's *In Memory Yet Green* and Pohl's *The Way The Future Was*, and Damon Knight's memoir *The Futurians* includes funny portrayals of them both. During those years they wrote a few stories together, as most of the members of that group had a habit of doing. Quickly Asimov's solo stories established him as one of the most famous science fiction writers alive. Pohl's career had a more scattered beginning, and along with writing short fiction, often in collaboration with Cyril Kornbluth, he worked extensively as an editor and a literary agent. Both men joined the Army during World War II, and Pohl worked as an Army meteorologist, giving him experiences that are perhaps relevant to this book. I recall a dinner in Portales, New Mexico, in 2005, during which Pohl discussed the phenomenon of virga with pleasure; it's a kind of rainfall that doesn't reach the ground, and he liked both the word and the sight of those sheets of rain hanging in the sky.

Soon after the war ended both men returned to civilian life, and wrote prolifically through the next four decades. Though somewhat similar in their aesthetics and politics, they were quite different in character, at least in public. Asimov was sanguine and ebullient, a cheerful polymath and public intellectual who could write well on any topic, and spent most of his time in Manhattan. Pohl was saturnine and watchful, and was mostly a science fiction insider, who traveled extensively and lectured frequently. They both remained as committed to liberal politics as they had been in their youth, which put them at odds with some other writers in the science fiction community, especially during the Vietnam War. They were both forthright advocates of the scientific method and the scientific community, and both could be called environmentalists from the moment the term was invented.

By the end of the 1980s, both of them had become concerned that the multiple ecological problems afflicting the planet were going to merge into a larger biosphere crisis that would be greatly exacerbated by climate change. What could they do about it? They were writers, and so they concluded their best chance of making an effective intervention was to write a book warning their fellow citizens of the danger of the situation. This book would first list and analyze all the problems, then suggest viable solutions to them. Writing such a book wouldn't be easy—the subject is massive and complicated—but these two writers were up to the task. Asimov was simply amazing in his ability to comprehend and synthesize large bodies of information and present them clearly. The four hundred books he wrote in his lifetime covered nearly that many topics, and these weren't mere summarizations; his clarity of expression, good judgment, and awareness of context made them truly interesting. Asimov on the Bible? On Shakespeare? On chemistry? He was really good on all these subjects, and many more. As for Pohl, he was also a polymath and a fine writer, and the environmental situation was his special topic of expertise, and a matter of increasing concern to him.

So the two men made a great team, and they combined to write in a clear informative style. You can't distinguish their voices in this text, but I suspect they split the job about equally. Even if the book was Pohl's idea, and he talked Asimov into lending his fame to the project by coauthoring it—even if Pohl brought the bulk of the preparatory materials to the table—I think it most likely Asimov jumped fully into the work, and did his fair share or more. Write half of a synthesis of the entire planetary situation? This was what he liked to do and was always doing. So here these two old pros took on a task they cared deeply about, and the result is very impressive.

Kim Stanley Robinson
January 2017

INTRODUCTION

Isaac Asimov

Throughout history, the doom-criers have been with us. We have heard of Cassandra, the daughter of Priam of Troy, who told the Trojans constantly that their city would be destroyed—but was never believed.

Before her, there must undoubtedly have been prophets of doom among the Egyptians and the Babylonians, and the history of the early Jews was particularly rife with such matters. The prophet, Jeremiah, was constantly predicting the destruction of Judah, and after him there came a long line of people (including John the Baptist), who said, "Repent, for the Kingdom of Heaven is at hand."

The Day of Judgment (which represents the coming of the Kingdom of Heaven) is an ever living threat, and even today, such sects as Jehovah's Witnesses and the Seventh-Day Adventists expect it at any time.

All these doom-criers, however, based their hopeless commentary on religion. Humanity was full of sin (by which most religious people meant "sex," for they never seemed as concerned about murder, theft, and corruption as they did about a little sexual amusement) and, as a result, a righteous and vengeful god was going to destroy everyone and everything. Look at the Flood. Look at Sodom and Gomorrah.

Few people ever took these religious doom-criers seriously, however, for the simple reason that few people ever agreed on religion, and because thousands of years of threats of divine retribution had always come to nothing, anyway.

—But now the situation has changed.

It is not adultery and fornication that is threatening humanity, but physical pollution. It is not an angry god who is threatening to destroy everything; it is a poisoned planet—poisoned by us.

Humanity is being threatened by its own deeds, yes, but the deeds that are threatening us with destruction do not involve the breaking of the ten commandments.

The coming of doom is, instead, the result of deeds that do not seem evil on the face of it. Because we are concerned with improving the health of mankind, and its security, our population has increased markedly, especially in the last hundred years, to the point where Earth cannot support us all.

Because we have industrialized ourselves in order to lift the curse of physical labor from our backs, we have poured the poisons produced by the internal-combustion engine into our atmosphere and dirtied it to the point where we can scarcely breathe it.

Because we have learned to make new materials for the greater convenience of mankind, we have produced chemical toxins that are saturating our soil and water.

Because we have found a new source of energy (and destruction) in the atomic nucleus, we face the threat of nuclear war, or even if we avoid that, the permeation of our environment with dangerous radiation and nuclear wastes.

This book is not an opinion piece. It is a scientific survey of the situation that threatens us all—and it says what we can do to mitigate the situation.

It is not a hopeless cry of doom at all. It is a description of what we face and what we can do about it. And in that sense, it is a hopeful book, and should be read as such.

It is not too late—

But it may become too late, if we wait too much longer.

Isaac Asimov

INTRODUCTION
Frederik Pohl

Let me tell you why we thought it was so important that this book be written, and what it is meant to do.

There have been many recent books on the environment, and how we are abusing it, many of them excellent. Among them they have laid out the vast variety of ways in which the activities of people like ourselves are damaging the health of our planet. Some of these books have even told us what each one of us can (and should) do in our daily lives to slow down the rate of destruction—by recycling, by refusing to buy the most destructive products, by arranging our lives so as to use everything we need more efficiently and thus to need less.

All of that sort of information is certainly very important, and in case you've missed any of it we'll take time to tell you again here.

But if every one of us does all those things it still will not be enough.

It is already too late to save our planet from harm. Too much has happened already: farms have turned into deserts, forests have been clear-cut to wasteland, lakes have been poisoned, the air is filled with harmful gases. It is even too late to save ourselves from the effects of other harmful processes, for they have already been set in motion, and will inevitably take their course. The global temperature will rise. The ozone layer will continue to fray. Pollution will sicken or kill more and more living creatures. All those things have already gone so far that they must now inevitably get worse before they can get better.

The only choice left to us is to decide how *much* worse we are willing to let things get.

We still have time to save, or restore, a large part of the gentle and benevolent environment that has made our lives possible. We can't, however, do it easily. We can't do it at all without at the same time making considerable social, economic and political changes in our world. These changes go far beyond anything we can accomplish as individuals; and to describe why these large-scale changes are necessary, what they must be, and how we can make them happen, is what this book is about.

Let me give you a sort of road map to the book, so you can know what to expect.

First we will start with a sort of overview on how to think about the environment. We'll look at the recent highly environmental war in the Persian Gulf. Then we'll look into such matters as "Gaia" and other hopes; what conservation requires us to conserve; how much we can believe about future projections and so on.

In the next part we will examine the major environmental threats to the world we live in, and what kinds of damage they will do to us if we let them. There isn't any good news in this section. If you've already learned a good deal about global warming, acid rain, and all the other threats, much of what this part of the book contains may not be news at all;

but in it we will try to explain all the processes involved in layman's terms, as well as to give you enough information to let you decide for yourself how real the threats are.

Next we will come to some reasonably good news. There are plenty of technological alternatives to our present machines, power plants, energy sources, etc., as well as alternatives in our daily lives. Here we will see how we can use them to do things better and still maintain a comfortable standard of living.

Then we will begin to cover some quite new ground, first by looking into the social and economic changes that our environmental problems will bring about.

There isn't any real doubt that major changes are inevitable. The only question is what kind. Some of them will happen no matter what we do, because as the environment deteriorates they will happen automatically. Others will be brought about by our efforts to stave off disaster. All the changes will be significant, and the world of the next generation is going to be quite different from our own.

Finally we will get into the political aspect of true conservation: why real change will be difficult, and what political actions we can take to bring it about.

I know that this part isn't good news, either. To ask the average self-respecting American to take an active part in the notoriously dirty business of politics is not unlike asking him to consider entering a career in street prostitution. But if we want to prevent the worst of the disasters we have no alternative to political action. Individuals can't do the job on their own; it's too big. Only government action can carry through the changes that must be made. And governments are both created and controlled by politics.

* * *

I almost feel I have to apologize for making you work so hard.

I have had this feeling before. I have spent a good deal of time in talking about the dangers to the environment long before they became a fashionable subject—for more than thirty years now. Sometimes I've done it in the books I've written, sometimes in the course of my part-time career as a lecturer, going around the world to give speeches to groups of all kinds. Over the years I must have given a couple of thousand talks and, although they have been on many subjects, I usually have worked the environmental questions into them somewhere.

Generally speaking, the audiences I've addressed have been made up of caring, intelligent people—rather like what I imagine the readers of this book to be, in fact. And yet, in every talk, somewhere along the course of the recital of approaching disasters I've become aware of a sort of stillness that comes over the audience. The people listening are always polite. They're even attentive. All the same, I can see that they are also beginning to wish strongly that the catalogue of bad news would end pretty soon.

I sympathize with all those people. I wish it could end, too.

The trouble is that things haven't got better over that third of a century. True, there have been a handful of real victories. A few lakes are cleaner than they were. Even in downtown Pittsburgh you can sometimes see a star or two in the nighttime sky. In New York's East River an astonished angler caught an actual live fish not long ago. The United States has banned the use of ozone-destroying CFCs in spray cans (though not their manufacture and use in other ways).

But all of these partial triumphs are not enough. For every gain there have been a dozen losses; as we will see, taken all in all, our world is dirtier and more threatened now than it has ever been in the past, and it is sure to go on becoming more and more so if we let it.

I sympathize in another way, too. Like most of my listeners, I sometimes find it hard to believe in my heart that all these large-scale environmental problems have much to do with *me*.

They don't, after all, look very real—yet. I know just as well as my listeners do that, when I wake up tomorrow morning and look out my window, things won't look so bad. The sun will still be shining. The trees in my backyard will still be green. There will still be food in the supermarkets and no one will be staggering down the streets, blinded by ultra-violet radiation. There isn't any doubt that some terrible things are in the process of happening to our world. But the worst of them haven't happened *yet*.

So, really, why should any of us get very excited now about calamities that may be decades away?

I do get excited, though.

I have seven very good reasons for serious concern. Their names are Christine, Daniel, Emily, Eric, Julia, Tommy and Tobias. They are my grandchildren.

As I write this they range from the very small to the middle teens, and when they grow up and have children of their own I would like very much for them, too, to have green trees around them and plenty to eat, and for them to be able to walk in the sun without fear of a nasty death, and to know that the world will survive.

It looks very much as though they may not have all of those things, though. They are lucky enough to have been born to a great competitive advantage: They all live in parts of the world that will be among the slowest to suffer from what we are doing to it. But that will not save them for long . . . not unless you and I and a lot of other people do get excited, excited enough to do *now* the things that will give them their birthright of a good life *then*.

It can happen. There can be a happy ending to it all . . . if we have the wisdom and the willingness to make it happen, by doing some pretty difficult things.

If we don't, there won't be any happy ending. There will merely be—for many of the things that make the world good, and very possibly for much of the living world itself—an ending.

Frederik Pohl

THE
BACKGROUND

1

The Environmental War

When nations go to war each side tries to destroy the other side's fighting forces—or, sometimes, just the other nation; that is what war is about.

But there is a third party in every war. It commits no hostile act against either side. Nevertheless, it is attacked by the bombs, missiles and cannon of both. That is the environment. When the war is over one combatant side or the other may claim victory. But the environment always loses.

The Persian Gulf War of 1991 was a totally environmental war. Even the causes which brought it about were environmental, for they began with a struggle over the fossil fuel, oil, which is at the root of so many environmental evils. The Persian Gulf War produced the same crop of environmental disasters as all wars do, as burning cities released their toxins into the air, and destroyed water and sewage systems produced their crops of sickness, and land areas and shorelines were sown with mines, and the debris of great armies littered the battleground. But it also produced a number of environmental disasters that were not usual at all. They came from oil. Most of the world's known reserves of oil lie under the sands of the Persian Gulf. Some of those desert lands are hardly more than a thin skin over underground oceans of petroleum. As defeat approached for Saddam Hussein, he tried to postpone it by using the oil itself as an environmental weapon. He ordered his troops to open the valves so that vast stocks of that oil poured into the sea, and to set all the oil wells and storage tanks of Kuwait afire.

Iraq wasn't the first nation to use environmental weapons against its enemies. Others have done it—even the United States; that is what the American forces did when they sent aircraft to spray plant-killing chemicals like Agent Orange on the farms, rice paddies and forests of Vietnam. If Saddam Hussein's action was morally worse, it is only because it was more pointless. These actions couldn't win the war for him. They couldn't even prevent his total defeat. They could never be more than a brutal act of revenge on his enemies. And in the event, they harmed his own people as well as his enemies . . . and a large section of the nearby world.

The Persian Gulf War is the war oil made. Without oil there would hardly have been any armies on the scene that were capable of fighting it.

By August 2, 1990, when the Iraqi army invaded its little neighbor, Kuwait, the nations of the Gulf already possessed some of the world's most powerfully armed military forces. Those armies were expensive. About a quarter of Iraq's large Gross National Product was routinely spent on arms, and Iraq was not alone. Saudi Arabia spent almost as high a proportion—20%—of its own, much larger GNP, and the other states of the area were not far behind.

It was oil that armed those countries. The cost of every last machine-gun cartridge and combat boot was paid for out of the money the industrialized nations paid to the nations of the Persian Gulf for their oil. Without oil, they would all be poverty-stricken. Before oil, they had little to sell in the world market but fish and pearls (and even the pearls were a dwindling source of revenue as Japanese cultured pearls took away their buyers). With oil—and with our insatiable appetite for burning the stuff in our cars, homes and industries— they take in the vast flood of foreign currency with which they buy the French Mirage fighters and American AWACs and Soviet T-72 tanks. If we didn't waste such prodigious quantities of energy—to the ecological harm of the world, as well as to ourselves—we wouldn't have to buy their oil. In that case none of them could afford their immense investments in the machineries of killing. It's as simple as that.

Not much else about the Gulf War, its causes and its effects, is very simple, though. The reasons why President George Bush reacted with such rage and forceful action are particularly complex. Though he specifically denied it, they certainly had to do with fears for the loss of American sources of imported oil, but they also had to do with the history of American involvement in the area, especially our complicated, seesawing relations with the neighboring country of Iran.

American involvement with Iran also began with oil. When, decades ago, the prime minister of Iran, Mohammad Mossadegh, nationalized Iran's oil industry, the American CIA sent their agent Kim Roosevelt in to arrange his overthrow. The revolt succeeded; the United States restored the Shah to the throne he had been expelled from, and in payment for services rendered the Shah's new government gave a 40% share of Iran's oil to American companies.

Of all America's purchased allies, the Shah of Iran was perhaps the most loyal. He stayed bought. In commerce, Iranian oil continued to enrich American companies; in foreign affairs, Iranian policy supported America without question; in its own territory, Iran granted America military bases and listening stations and missile emplacements all along its long border with the Soviet Union.

The Shah, however, was not a democrat. The oil wealth that flowed into Iran's economy enriched only a tiny fraction of its people—the fraction closest to the Shah. The Shah's regime created no significant middle class, and the bulk of the Iranian population not only remained poor but suffered considerable tyranny. Americans on almost every college campus in the United States during the last years of the Shah's reign became familiar with the spectacle of Iranian students, wearing paper bags over their heads to prevent retaliation against their families back in Iran, protesting the brutality and torture of the Shah's American-trained secret police. The brutality and poverty were real. Their result was revolution, the deposition of the Shah and the delivery of Iran to the Ayatollah Khomeini. Khomeini, not entirely unjustly, blamed the United States for the worst of the Shah's policies. Inflamed by Khomeini's oratory, his Revolutionary Guards sacked the American embassy and took every American diplomat they could catch hostage—the embarrassment that crippled the final years of Jimmy Carter's presidency, and played a large part in the election of Ronald Reagan.

The result was that, in the eyes of the American government, Iran changed overnight from close ally to hated foe. And the result of that was that when, a decade ago, Saddam Hussein massed his army and invaded Iran, the United States did not interfere by word or act. There was no talk of the wickedness of unprovoked aggression then. Indeed, when at last the United States did take action, it was on the wicked aggressor's side.

Curiously, the precipitating incident was the attack on an American naval vessel by one of the wicked aggressor's own planes. An Iraqi aircraft fired a missile at the U.S.S. *Stark*, severely damaging the ship and killing 47 American sailors. The response of the Reagan administration was two-fold. Overtly, it dispatched additional American warships to the Gulf. Covertly, it began passing on to the Iraqis satellite information on the disposition of Iranian troops, warning of buildups that threatened surprise attacks (and saving Iraq at least one or two damaging defeats); winking at the export to Iraq of some high-tech computer and other militarily useful devices which had previously been denied them; and helping Iraq's economy by providing it with food from America's farms—making Iraq the second most favored nation in the world, after only Mexico, as the recipient of such largesse.

As a result, it appears that President Ronald Reagan, and President George Bush after him, concluded that they had bought themselves another ally in Saddam Hussein. Certainly they did their best to protect his government, even against the Congress of the United States. Right up to the very eve of Saddam Hussein's invasion of the tiny (but oil-soaked) country of Kuwait, the Bush administration was sending its arm-twisters to Capitol Hill to try to prevent any vote in Congress to impose sanctions on Iraq for its human-rights violations. Just days before the invasion, America's ambassador, April Glaspie, specifically notified Saddam

Hussein in a face-to-face conversation that the United States took no position on his territorial disputes with Kuwait. Presidential administrations speak often of the "signals" they send to other nations; in this case, the signal could only have been read as "go."

Why then did George Bush respond so quickly and violently with outright war?

It was not because the nation compelled him to it. The public-opinion polls showed that a large majority of Americans wanted restraint. In testimony before Congress, nearly every important American involved in foreign policy counseled the same. Perhaps a majority of his own administration advised going slow—yet Bush plunged ahead. His emissaries covered the globe, buying up new allies—$14 billion in forgiven debts to Egypt, a billion to Syria, $9 billion's worth of arms to Turkey, $4 billion to the USSR (but Saudi Arabian money, this time), even $115 million and a cordial Washington reception for the quickly forgiven (for the Tiananmen Square deaths) People's Republic of China. The Bush administration even did its best to demand that the governments of Japan and Germany, forbidden (by measures the victorious United States had originally imposed on their defeated governments after World War II) ever to send their troops against foreign nations, now break their own laws and join in the fighting war against Iraq—unsuccessfully as far as troops were concerned, but securing cash contributions to the cause at least. And when all the United Nations resolutions had been passed and a quarter of a million American troops were lining the Saudi border and the embargo was beginning to take effect, George Bush sent another quarter-million troops to the area, declared sanctions weren't working and began the air war that reduced Iraq to Third World status before the ground forces invaded.

Why?

The only answer lies in the psychological makeup of George Bush. He had tasted the joys of the conqueror a year earlier, when American forces invaded Panama and put a new government in place; he remembered the "wimp" label that had been placed on him; he felt personally betrayed by a client who had committed an unauthorized aggression; all of these were undoubtedly factors in his decision.

Beyond that, there was one other element in the makeup of George Bush. He was, after all, a Texas oilman. And not just in Texas; his concerns included the oil of the Persian Gulf. Kuwait's very first offshore oil well had been drilled by Bush's company; even during the war, his eldest son's company was exploring for oil elsewhere in the Gulf. Whatever else motivated President Bush, oil drenched all his acts.

And oil, finally, was the weapon that Saddam Hussein used to strike back at his enemies when his own defeat was inescapable. He flooded the Gulf with oil runoff; he set Kuwait's oil wells on fire.

It was the pouring of crude oil into the Persian Gulf that did the worst long-term environmental damage. How long that long term will turn out to be is still a question; the Gulf is vulnerable in many ways.

The Persian Gulf is a shallow body of water with wide mudflats and beaches on its shores. Along much of the Kuwaiti and Saudi Arabian shoreline the distance between low- and high-tide marks is often a mile, sometimes as much as six miles. These crucial miles are where the grassy salt marshes and other "wetlands" are found, the places where most sea life begins. Great stretches of these intertidal wetlands are already heavily damaged by the demands of the oil industry— dredged to create loading facilities for tankers or filled to make room for oil storage tanks—but the biologically active areas which remain are important to the life of the whole Gulf.

All these wetlands, along with the shallows just offshore, are highly fertile. Like similar regions all around the world, they produce large quantities of algae and other organisms on which birds, fish, crustaceans and marine mammals feed. Higher in the food chain, people live on the results of this productivity; the fisheries of the Persian Gulf produce about 150,000 tons of edible shrimp, bivalves and finfish each year. And, of course, the algae of the Persian Gulf are continually performing the essential life-support task that all plants all over the world are constantly doing for us: they take carbon dioxide out of the air and replace it with the oxygen we need to breathe.

When the spilled Persian Gulf oil hits one of these broad, flat beaches it stays. As the tide comes in, it brings the oil with it; when it retreats, the oil stays behind. There are ways of cleaning oiled beaches. They are not very successful; two years after the much smaller *Exxon Valdez* spill there is still oil under the surface of most of the polluted areas. (Exactly how serious the remaining *Exxon Valdez* pollution is has not been made public, even two years after the disaster. Much of the data gathered by investigators has been impounded for use as evidence in the large number of ongoing lawsuits, some of which may continue for years to come.) Along many of the

Kuwaiti and Saudi Arabian beaches the situation is far worse than in Alaska, since the Gulf beaches have been sown with mines to deter enemy landings; the cleanup workers don't have only the problems of any such task, they must also face the risk of having a foot blown off in the process.

Oil pollution is nothing new to the Persian Gulf. An earlier great oil slick was released during the Iran-Iraq war; attempts were made to deal with it, but what ultimately happened was that it simply disappeared. (Experts think the remains of the oil simply absorbed enough water to be heavy enough to sink, and now foul the bottom of the Gulf at some unknown location.) Smaller spills happen all the time. On average, one reportable spill happens every month, year in and year out; a majority of the Saudi beaches are chronically afflicted with oil residues. But the two great spills of the 1991 Gulf War are by far the worst. At least half a million barrels of oil, perhaps five times that, formed a miles-long slick that slowly drifted across the Gulf, killing as it went. No one has counted the numbers of turtles, cormorants, dugongs, shellfish, dolphins, finfish and other living creatures that have been killed directly by the oil, or those that will die of starvation because of the loss of their food supplies.

Oil floats longer in the Persian Gulf than in most seas because of the Gulf's high salinity, in some places nearly twice the world average. Still, sooner or later the waters of the Persian Gulf will clean themselves—if they are allowed to; if new oil spills don't perpetuate the pollution. But that time will be years in coming, and meanwhile the ecology of the whole area has suffered a serious blow.

* * *

On the other hand, it is Kuwait's five hundred flaming oil wells that did the most immediate damage to human beings.

This was the biggest sustained fire the world has ever seen, somewhere around five million barrels of crude oil burning every day. When a single well goes up in flame it's a terrifying spectacle; the people who make a business of putting out oil-well fires are highly trained specialists, and their lives are at risk. Five hundred such fires burning at once is something new in the world. Just quenching the flames is not enough; if another well is afire nearby, it may reignite the one just controlled. Nor is that the worst. As with all post-battle attempts at cleanup, the work of trying to put out burning Kuwait is made far more dangerous because the Iraqi troops often mined the area around each well; in addition to which, some of the explosive charges are known not to have gone off, and their presence at the wellheads is a continuing threat.

So snuffing out the flame—say, by setting off explosives nearby—may not do the job. Extinguishing it by depriving it of the oxygen it needs to burn—by erecting a collar around the wellhead and filling the collar with nitrogen or carbon dioxide—is difficult at best, here doubly so because of the risk of mines. For some of the burning wells, the only solution may be to drill another well alongside the burning one, right into the shaft of the well on fire, and filling the relief well with cement or heavy liquids to shut off the flow of oil to the surface—a task that may require a month or more of hard, expensive work for each well.

And there are five hundred of them.

As they burn their millions of barrels each day, they produce immense clouds of sooty smoke. The smoke shades the earth below—temperatures have been measured almost twenty degrees colder under the smoke pall—and the soot gets in people's lungs, settles on farmlands and contaminates them and falls on the waters of the Gulf, adding to the pollution from the oil spills. Because Kuwaiti oil is "sour"—heavily laced with sulfur—the fallout is a particularly poisonous form of acid rain, and when the rain falls it is sometimes black and oily. (Some observers say that it looks like the rains that were reported to fall near Hiroshima after the Japanese city was atom-bombed in World War II.) People in Kuwait City reported "darkness at noon" from the smoke pall; visibility at the airports hampered takeoffs and landings; American troops wore masks to filter out the worst of the soot, and so did local residents. The harm to respiratory systems, particularly of children, is unknown but undoubtedly severe.

Middle Eastern countries have threatened to wage environmental war before. Twice—in 1978, and again in 1985—Egyptian officials have explicitly threatened war against Ethiopia and Sudan if those nations interfered with plans to construct dams which would interrupt the flow of the sources of the Nile. Even in the present conflict, Turkey threatened to starve Iraq of water by cutting off the flow of the Euphrates River (and might have done it, except for the fact that the Euphrates serves Syria as well as Iraq, and in this particular confrontation Turkey and Syria were on the same side).

But it was Saddam Hussein who wielded the environmental weapon in the Persian Gulf; and the whole area will be a long time recovering from its use.

What the environmental war has done to the Persian Gulf is only a closeup example of what we are doing to the whole world. The war is over now; the rest of what we are doing goes on. The five million barrels of oil that burn from the wells of Kuwait every day amount to less than a twelfth of the 66 million barrels we burn every day, worldwide, in our cars, industries and homes.

Can we hope that the world, too, will somehow recover from the environmental harm we do?

Perhaps we can. Indeed, there is a name that has recently been given to this hope for Earth. It is called "Gaia." In our next chapter we will try to see how much we can rely on Gaia (and some other hopes) to solve our environmental problems for us.

2

Gaia and Other Hopes

We would all like to think that there was something—some benign and superior kind of *Something*—that would step in and save us from the things that are going wrong with our world.

Most people have always had a comforting belief of that sort. In most of human history their nominee for that "something" was usually their god—whatever god they chanced to worship in that time and place—which is why, in parched summers, farmers have long prayed for rain. They still do, but as scientific knowledge grew and began to explain more and more events as the working out of natural law, rather than divine caprice, many people began to wish for a less supernatural (and perhaps a more predictable) protector.

For that reason there was quite a stir in the scientific community when, about twenty years ago, an English scientist named James Lovelock came up with something that came close to filling that bill. Lovelock gave his hypothetical new concept a name. He called it "Gaia," after the ancient goddess of the Earth.

When Lovelock published his "Gaia hypothesis" it shook up many scientists, especially the most rational-minded ones, who purely hated so mystical-sounding a concept. It was an embarrassment to them, and the most disturbing part of it was that Lovelock was one of their own number. He did have a reputation as a bit of a maverick, but his scientific credentials were solid. Among other accomplishments, Lovelock was known as the scientist who had designed the instruments for some of the life-seeking experiments that the American spacecraft *Viking* carried out on the surface of Mars.

Yet, in the eyes of his peers, the things that Lovelock was saying verged on the superstitious. Worse, he had the temerity to present his arguments in the form of the orthodox "scientific method." He had drawn the evidence for his proposal from observation and from the scientific literature, as a scientist is supposed to do. That evidence, he said, appeared to show that the entire biosphere of the planet Earth—which is to say, every last living thing that inhabits our planet, from the bacteria to the whales, the elephants, the redwood trees and you and me—could usefully be described as one single, planet-wide organism, each part of it almost as related and interdependent as the cells of our body. Lovelock felt that this collective super-being deserved a name of its own. Lacking inspiration, he turned to his neighbor, William Golding (author of *The Lord of the Flies*) for help, and Golding came up with the perfect answer. So they called it "Gaia."

Lovelock came to his conclusions in the course of his scientific work while he was trying to figure out what signs of life the instruments he was designing should look for on the planet Mars. It occurred to him that if he had been a Martian instead of an Englishman, it would have been easy to solve that problem the other way around. All a Martian would need was a modest telescope, with a fairly good spectroscope attached, to get the answer. The very composition of Earth's air proclaimed the undeniable existence of life. Earth's atmosphere contains a great deal of free oxygen, which is a very active chemical. The fact that it was free in such volume in the Earth's atmosphere meant that something had to be constantly replenishing it. If that were not true, the atmospheric oxygen would long since have reacted with such other elements as the iron in the Earth's surface and thus have disappeared—in just the same way that our own Earthly spectroscopes have shown that whatever oxygen there ever was has long since been used up on all of our planetary neighbors, Mars included.

Therefore a Martian astronomer would have understood at once that that constantly oxygen-replenishing "something" could be only one thing. It had to be life.

It is life—living plants—which continually produces that oxygen in our air; it is that same oxygen which life—ourselves and almost every other living member of the animal kingdom—relies on to survive.

Lovelock's insight from that was that life—all terrestrial life combined—was interactive and had the capacity to maintain its environment in such a way that its own continued existence was possible. If some environmental change should threaten life, life would then act to counter the change, in much the same way that a thermostat acts to keep your home comfortable when the weather changes, by turning on the furnace or the air conditioner.

The technical term for this kind of behavior is "homeostatic." According to Lovelock, "Gaia"—the sum total of all life on Earth—is a homeostatic system. (To be more technically precise, the proper term is "homeorhetic" rather than "homeostatic" in this case, but the distinction would matter only to specialists.) This self-preserving system not only adapts itself to changes, it even makes changes of its own, altering the environment around it in whatever ways are necessary to its well-being.

With that speculation to spur him on, Lovelock began looking for other evidences of homeostatic behavior. He found them in surprising places.

Coral islands, for instance. Coral is made up of living animals. They can only grow in shallow water. Yet many coral islands are slowly sinking, and somehow the coral continues to grow upward just as much as it needs to remain at the proper depth for survival; that's a rudimentary kind of homeostasis.

Then there's the temperature of the Earth. The global average temperature has stayed within fairly narrow limits for a billion years or more, although it is known that in that time the radiation from the sun (which is what basically determines that temperature) has been steadily increasing. Therefore the Earth should have warmed appreciably. It didn't. How could that happen without some kind of homeostasis?

Even more interesting to Lovelock was the paradoxical question of how much salt there is in the sea. The present salt concentration in the world's oceans is just about right for marine plants and animals to live in. Any significant increase would be disastrous. Fish (and other sea life) have a tough job as it is in preventing that salt from accumulating in their tissues and poisoning them; if there were much more salt in the sea than there is, the job would be impossible and they would die.

Yet, by all normal scientific logic, the seas should be a lot saltier than they are. It is known that the rivers of the Earth are continually dissolving salts out of the soil they run through and carrying more and more of those salts into the seas. The water itself which the rivers add each year doesn't stay in the ocean. That pure water is taken out by evaporation by the heat of the Sun to make clouds and ultimately to fall again as rain; while the salts those waters contained have nowhere to go and must stay behind.

We know from everyday experience what happens in that case. If we leave a bucket of salt water exposed in the summer, it will get more and more salty as the water evaporates. Astonishingly, that does not happen in the oceans. Their salt content is known to have remained at just about the same level for all of geological time.

So it is apparent that *something* is acting to remove excess salt from the ocean.

There is one known process that might account for it. Now and then, bays and shallow arms of the ocean are cut off. The Sun evaporates their water and they dry out to form salt beds—which ultimately are covered over by dust, clay and finally impenetrable rock, so that when the sea ultimately returns to reclaim that area that layer of fossil salt is sealed in and is not redissolved. (When, later on, people dig them up for their own needs we call them salt mines.) In that way, millennium after millennium, the oceans get rid of the excess and keep their salt content level.

It could be simply a coincidence that that balance is maintained so exactly, no matter what else is going on . . . but it could also be another manifestation of Gaia.

But perhaps Gaia shows herself most clearly in the way

she has kept Earth's temperature constant. As we've mentioned, in the early days of Earth the Sun's radiation was about a fifth less than it is now. With so little warming sunlight the oceans should have frozen over, but that didn't happen.

Why not?

The reason is that then the Earth's atmosphere contained more carbon dioxide than it does now. And there, Lovelock says, Gaia is at work. For plants came along to reduce the proportion of carbon dioxide in the air. As the Sun warmed up, the carbon dioxide, with its heat-retaining qualities, diminished—in exact step, over the millennia. Gaia worked through the plants (Lovelock suggests) to keep the world at the optimum temperature for life.

Is Gaia real?

A number of scientists have come to believe so to one degree or another. Some of them cite the fact (as evolutionary studies seem to show) that Earth's living creatures are all cousins, every one of them descended from the same few primitive ancestral organisms that existed more than three billion years ago. Since the living things on Earth are all related it is not unreasonable to think that they are interdependent in many subtle ways.

Nevertheless, most scientists will have none of Lovelock's idea. But whatever the truth of the Gaia hypothesis it is an unquestionable fact that Gaia (if such a thing really does exist) is under heavy attack today. Thousands of species of the living things which make up the parts of Gaia have already been exterminated (the passenger pigeon, the auk, the smallpox organism) and many thousands more are likely to disappear in our own lifetimes (the mountain gorilla, the white rhinoceros, the blue whale).

Extinctions of species are nothing new in the long history of life on Earth. Over the eons since the first living things appeared there have been any number of them. Species have become extinct because of climate changes, or because more efficient living things appeared and wiped them out, perhaps at least once because a giant asteroid struck the Earth and blotted out the Sun for a year or more in the great dying-off that occurred at the end of the Cretaceous period. Since the first appearance of life on Earth, hundreds of millions of species have disappeared, an average of something like one species going extinct every year. "Gaia," however, still survives. As Lovelock himself puts it, Gaia is like a tree: 99% of it, bark and wood, is dead, but the tree itself is still alive and well.

The worrisome fact for us to consider is that the current waves of extinction of species are not like the great die-offs of the past.

Species extinction now is happening much faster—several hundred or even several thousand species dying off each year instead of one—and the causes of extinction are different. Now it is not any natural accident that is killing most of the species that are disappearing. What is killing them is human beings. Sometimes we do it by hunting them down, sometimes by introducing predators against which they have no defense. More often we exterminate them simply by destroying the environment in which they live.

The great new worry is that now it begins to seem possible that we ourselves are threatened with the same fate.

No matter what happens, we won't all die tomorrow, of

course. With a little luck, we won't all die—at least, the probabilities are very high that not every member of our human species will die—for a good long time to come.

But we can no longer be quite positive of that. Now we need the luck, since the things we can be quite positive about include the fact that we are in serious danger, perhaps quite soon and certainly sooner or later, from the consequences of one or all of such ongoing phenomena as acid rain, the destruction of the ozone layer, the global "greenhouse" warming and a dozen other current human interventions in our world.

Please remember that all of the unpleasant environment-destroying processes we've been talking about are unquestionably real.

That is not to say that there are no questions at all about them. There are a great many questions. They turn up in every day's newspapers. But the questions, almost without exception, turn only on how far these processes will go, and whether they can already be observed in action, and what their long-term consequences will be. There is very little room for doubt about whether the processes themselves are real.

These processes are not only potentially dangerous to human well-being and even to human life, they all have one strange and bitter quality in common.

The human race, however fearfully it has faced predators, weather extremes, disease and a chronic shortage of food over its history, has managed to survive all the threats from nature. Really, we've done a remarkably good job of combating the traditional killers of our race. We now make our own environments in our homes. We grow whatever we need to eat. No one dies of smallpox any more, because the organism that causes that disease has been made extinct. Fewer and fewer die of polio, cholera or any of the traditional killers. Wolves do not come into the streets of our cities and carry off our babies.

We still do die "natural deaths," of course, but most of even those "natural" deaths are different in kind. The "natural" things we die of now are more likely to be cancers and cardiovascular complaints—which are often exacerbated, and sometimes caused, by our diets, our habits and our man-made environment. Other significant mortality factors are car accidents, murders, suicides—and wars. We have in fact conquered so many natural threats that now there are only two things which stand any reasonable chance of seriously depleting the human population on Earth over the next few decades or so. One of them is nuclear war. The other is our destruction of the environment we depend on for life.

Having said this, we can see that what almost all these current threats to human life have in common is that they are what are called "anthropogenic" processes: they are things we do to ourselves.

It is a distressing fact that even such other "natural disasters" as storms, earthquakes and so on are not entirely natural any more. If we don't actually cause such catastrophes, then we surely contribute to making their effects a good deal worse. Probably no one would have died in the October 17, 1989, earthquake that struck the San Francisco area if they had been living in "a state of nature." What actually killed almost every one of the victims was man-made structures collapsing and crushing them, or trapping and burning them to death. Even the death and injury toll from such storms as Hurricane Hugo is at least partly anthropogenic. One reason for saying this is that it is quite likely that a global warming significantly increases the number and intensity of hurricanes and may have already begun to do so—it is a fact that Hugo was by some measures the worst hurricane on record. In any case it is certain that many of those deaths, too, would have been averted if the human-made structures and machineries had not existed to crush the casualties.

All in all, it is clear that the human race is now contributing very significantly to its own deaths and injuries. "We have met the enemy," said Pogo, "and he is us."

Gaia is not the only hope some of us have for believing that we can somehow avert the consequences of our environmental sins. Some would believe, for instance, that some sort of socialist revolution would solve our problems. One member of the French school of socialist environmentalists, Andre Gorz, says (in his book *Ecology As Politics*): "The ecological movement is not an end in itself, but a stage in a larger struggle."

In a certain sense, Gorz is undeniably right. No amount of environmental action can solve all our problems. It won't even affect some of the worst of them—crime, war, poverty, illiteracy, etc. The most it can do is to keep some of our problems from getting worse, and thus to give us a *chance* to deal with the others.

But Gorz's hope that some form of socialism can solve the environmental crisis along the way to a better world is no more than an act of faith. All the evidence points the other way. The socialist countries of the world are not environmentally better than the rest of the planet. Indeed, they are far worse.

If one were to argue that "true" socialism has never been tried, that evidence could be set aside. But, really, there isn't any reason to believe that the farmers in a state-run agricultural society would be any less inclined to overload their lands with pesticides and chemical fertilizers than the present capitalist ones, or that a worker-owned automobile factory would be more likely to put itself out of business by making only energy-efficient and durable cars.

On the other hand, the most partisan "capitalists" among us have equally ill-based hopes. When the thoroughly capitalist Bush administration unveiled its 1991 energy program, it said almost nothing about energy *conservation*. To burn only American oil would help the federal budget a little. But American oil produces as much pollution as any other, and American coal is more polluting than any oil.

Then there is the Zero Population Growth—"ZPG"—movement. No doubt the world population is exploding alarmingly. No doubt at some future time the number of living people will exceed the carrying capacity of this planet, and future growth will have to stop.

But there is also no doubt that if all the three or four billion people of the Third World disappeared overnight, leaving only the billion-plus of us who live in the developed countries, the world's environment would still be facing crisis. That would help save the elephants and the rain forests, perhaps, but it would leave us with all the problems of ozone depletion, global warming and acid rain intact. More important than the simple number of people alive in the world is what those people *do*. The cause of the immediate crisis isn't sheer numbers, it is the unrestrained and wasteful use of energy and resources.

Lovelock's Gaia hypothesis was not only embarrassing, it was actively infuriating to many scientists.

The scientists could not deny that there was some sort of science at its base. All the same, they could not help but deplore it for its overtones of supernatural religion. (Certainly it was taken in that way by a great many non-scientists. As the biologist W. Ford Doolittle said, "A lot of people who don't believe in science really like Gaia.") Gaia was *comforting*. It offered a kind of motherly reassurance, in places where any reassurance at all—at least for any person who took a hard look at what was happening to the air, water and soil of the world—was hard to find.

But we must not take too much personal reassurance from Lovelock's theory. The Gaia hypothesis does suggest that life is likely to continue, and even that many species will exist. However, there is nothing in the theory to say that the world will be safe for our grandchildren, because there is nothing in it which predicts that the assortment of species of living things which will survive the present assaults will necessarily include that particular species called Homo sapiens sapiens, which is ourselves.

James Lovelock said it himself one day, talking over a cup of coffee with the author of *The Hole in the Sky*, John Gribbin. As Lovelock put it, "People sometimes have the attitude that 'Gaia will look after us.' But that's wrong. If the concept means anything at all, Gaia will look after *herself*. And the best way for her to do that might well be to get rid of us."

3

Inventing the Future

It's hard for some people to work up any real concern about the environment, because most of the worst news about what's happening to our world is really about things that are *going* to happen.

Most of us are inclined to take any predictions of the future with a grain of salt. We have good reasons for being skeptical, too. We all remember that terrifying predictions of disaster and doom have been coming at us for thousands of years. Some of them have come true. Many others (especially the worst of them, such as all-out nuclear war) simply have not.

So it is quite reasonable for any of us to ask, "Why—*this time*—should we believe these scare predictions will actually happen?"

We need to try to answer that question for a lot of reasons, not the least of which is that if you do believe what this book says you will probably want to do something to prevent some of these predicted events coming to pass . . . and that, as you will see, will involve you in a lot of things you probably hadn't ever thought of doing, some of which are hardly any fun at all.

So let's look at all the many ways in which useful predictions about the future can be made, in the hope that that can tell us something about what these dismaying environmental predictions are worth.

The scientific study of events that haven't happened yet is called "futurology." It is a respectable, at least fairly respectable, professional discipline, and it serves a clear and urgent need. Governments need to plan for disaster control, or for meeting growing needs for such services as water, waste disposal, transportation and many others. Corporations need to plan for expansion, model changes, new products, fluctuations in markets and so on. Hardly any human institution can conduct its affairs properly today unless it gives some thought to what it will be doing tomorrow.

Unfortunately, all too often the predictions are unreliable. Futurology is far from an exact science. Worse than that, any attempt at foreseeing the future contains some basic, built-in contradictions. The most important of these contradictions can be called the First Law of Futurology, and it can be expressed like this:

"The more complete and exact a prediction is, the less it is worth."

That seems to be against all common sense, but unfortunately it is true. The reasons why this is so become apparent when you observe that no prediction about tomorrow has any practical value unless you can use it to guide your actions today. It doesn't do you much good to know that something bad is going to happen if it is *sure* to happen, for if it is sure to happen there's nothing you can do about it.

Or, to put it in more formal terms: The major, if not the

only, utility of future studies lies in the ways in which their projections can help identify future problems, events or needs, so that, with the information the forecasts give us, we can do something *now* to bring about the desirable outcomes and try to avert the bad ones.

In any case, "complete and accurate" predictions do not exist, not even in theory. They can't. As Dennis Gabor (the scientist who invented the hologram, and one of the founding fathers of the discipline of futurology) put it:

"You can't actually predict the future at all. All you can do is invent it."

We can understand what Gabor means by this when we consider some modern history. For instance, it is fair to say that Adolf Hitler "invented" World War II and all its consequences. True, something like that war might well have happened even if Hitler had never been born, but it would not have happened at just that time or in just that exact way—and thus it, in turn, would have produced a different spectrum of later events.

Another example is President John F. Kennedy's "invention" of the manned lunar landings. Kennedy didn't create the idea of sending human beings to the Moon in rocket-ships. Many, many people predicted the *possibility* of space travel long before the first rocket ever left Earth's atmosphere. All the same, space travel did not become a *fact* until President John F. Kennedy, early in his term, seeking a project which would galvanize the American people (and insure him a place in the history books), put the power and prestige of his office behind the program which became the Apollo project. Nothing forced that decision on him; it was his own, and thus he "invented" that particular future.

What this book is about is the way in which we—all of us, the whole human race—are inventing through our actions a future which contains serious damage to our world . . . and the ways in which, if we choose, we can change course and invent a much happier one.

Now that we have become even more skeptical about future predictions than we were when we started, we can ask more pointedly than ever, How do we know that these warnings of disaster should be taken seriously?

Actually there's a good answer for that. A certain limited class of "future" events can be predicted very confidently . . . because they have already begun to happen.

As an example of that kind of "prediction," let's suppose that you (or your wife) are pregnant, and you would like to find out whether your unborn child is going to turn out to be a boy or a girl.

That's easily accomplished. Your doctor can insert a needle into the pregnant woman's body and withdraw a sample of fluid in the procedure called amniocentesis. When the fluid is analyzed in the laboratory, it can tell you the gender of the unborn child.

If the laboratory knows what it is doing, that prediction is very reliable. What it is not, however, is a "prediction." The child isn't *going* to be male or female. The child already *is* one or the other, but as it is still in the womb you can't yet see it for yourself.

The most troublesome "predictions" in this book are of that class: they describe things that have actually already begun to happen, though their full consequences haven't arrived yet.

We know pretty well some of the things which will occur if these processes are allowed to go on to their ultimate ends. What we don't know is whether the human race will take the hard steps necessary to halt or even reverse them somewhere along the line. That part of the future is still undecided. It hasn't been invented yet.

Like it or not, all of us are "inventing the future" every day, with all of our actions.

The capacity of any single individual to shape tomorrow's world may be quite small (unless, like John Kennedy, that person is a charismatic president or in some other way able to mobilize great resources for a major project). But there are a lot more of the rest of us than there are of charismatic world leaders. Collectively we five billion-plus living human beings are deciding by the everyday things all of us are doing in our ordinary lives today what the world of our grandchildren will be like . . . and some of those futures which we are now inventing are very worrying.

Indeed some of the present processes, if they are allowed to continue, will in fact be terribly damaging to a large part of the human race. There is even a real, if fortunately still quite small, possibility that some of them might wipe the human race out entirely.

4

Rationing Destruction

This planet that we live on seems pretty huge to most of us. It isn't easy for us to convince ourselves that any of the little wounds we habitually inflict on the Earth can make any real difference. Does it actually hurt anything if we jump in the car when we want to go a few blocks? Or drink out of foam plastic cups, or leave all the lights on when we go out? When a sport fisherman throws the plastic from a six-pack into the ocean, is it possible that he is doing any real harm?

Since each of us is so small, and the Earth so large, it's easy to think that the answer to all those questions is "no." Under the right circumstances, "no" would be the right answer, too—that is, if only one of us were doing them. What our common sense tells us is quite right: The damage any single individual can do to our habitat is trivial.

What makes it all non-trivial is that there are a *lot* of us, with more coming along all the time.

There will be four times as many people in the world at the end of the twentieth century as there were at the beginning, and collectively all those billions of us, with all our new machines, can do a great deal of damage. We *do* do a great deal. Eight per cent of all the carbon dioxide in the air today comes from the tailpipes of our cars. Those bits of plastic from a six-pack of beer, along with other plastic bits discarded from other vessels, have already become a significant cause of death among marine mammals. The five billion–plus of us who are alive on the Earth today collectively commit more environmental destruction every year than any war or any natural disaster ever has in all of human history.

When the Zero Population Growthers tell us that the fundamental pollution in the world today isn't carbon-dioxide pollution or chlorofluorocarbon pollution or acid rain, but people pollution, they aren't entirely wrong. There are simply too many of us, particularly in the developed world, to behave so badly.

It's important to understand that the problem is not simply that all these human beings are draining the carrying capacity of the Earth. People do make heavy demands on resources, with all their requirements for food, clothing, housing and other necessities of life, but that's not news. That particular problem was foreseen by Malthus long ago.

Malthus wasn't mistaken in worrying about future famines and the exhaustion of natural resources, he was simply ahead of his time. Sooner or later the human race must inevitably come to the point where Malthus's predictions come to pass, if it keeps on reproducing without limit. But we haven't reached that point yet, and meanwhile we have more urgent problems nearer at hand.

Actually, things have worsened since Malthus wrote. Simple scarcities no longer represent the real danger. Now the greatest threats to our grandchildren's future come from the fall-out from the things that we *do*, which is to say from the ways in which our heavily industrialized societies and profligate use of our resources are producing wastes and side-effects that are seriously damaging the world.

In the long run these wounds will largely heal themselves. Whether or not Gaia exists in any real sense, in the long run natural processes—there's no reason you shouldn't call them "Gaia," if you want to—will repossess the excess carbon dioxide from the atmosphere, wash out the acid from the rain, restore the ozone, nullify the toxic chemicals in the waters and the soil and repair all (well, almost all) the other damage that we have the ability to do to our planet . . . in the long run.

But, as John Maynard Keynes once said to President Franklin D. Roosevelt, "The trouble with the long run is that in the long run we are all dead."

These natural processes do their healing at their own pace, not ours. They can do the job of repair only at the rate fixed by natural law, which is not in our power to alter.

Unfortunately, that rate of repair is far slower than the rate at which we do the damage.

When we look at the problems in that way—as a race between damage and repair—we begin to see the glimmering of a place where we can look to find a permanent solution to our environmental problems. We can't speed up the natural rates of repair, but maybe we can slow down the rate at which we do the damage.

We can start by measuring all those natural rates of regeneration and repair. We can then go on to calculate roughly (for example) just how many gallons of gasoline, cubic feet of natural gas, tons of coal and other fossil fuels we human beings can afford to burn, worldwide, without adding carbon dioxide to the air faster than the world's forests can leach it out again—especially if, at the same time, we calculate how many trees we can afford to cut down every year, and how many new ones we must plant.

In the same way, we can make similar calculations for all of the other assaults we make against the environment. We can measure, for instance, the rate at which natural processes create new soil for cultivation (that happens at a rate of a fraction of an inch each year) and compare it with the rate at which erosion is now removing soil from our farms (a figure many times larger). We can add up the annual increase from rainfall and seepage to our underground water supplies, like the Ogalalla aquifer which supplies irrigation water for many Western states. We can then measure that against the rate at which dry-land farmers and cities are sucking the water out.

If we made such a calculation we could then divide the rate of inflow by the rate of withdrawal, and come out with a number. That number would be a valuable unit of measurement, of the kind that scientists call a "figure of merit." Such a figure could help us calculate just how much of all these things we can continue to do without destroying the natural balances.

We have a name for that proposed new unit of measurement. We would like to name it after ourselves—the species Homo sapiens sapiens—by calling it the "SAP."

If naming it in that way seems a little too facetious, the name can also be considered an acronym for "Steady-state Allowable Perturbations." But by whatever name we choose to call it, the SAP could be a very useful unit.

For example: If we know how many tons of ozone each year natural processes produce for the stratospheric layer that shields us from the most damaging ultraviolet radiation, we can then calculate how many tons per year we can afford to manufacture of the chlorofluorocarbons (CFCs for short) and other chemicals that destroy it.

Having worked that out, we then can use that SAP as a guide for our own individual behavior. We can simply divide the allowable tonnage by the total human population and thus find out how many computer boards or cans of CFC-spray deodorant, for example, we can each allow ourselves to consume in each calendar year.

If every human being alive then kept his own practices within those limits—not just for the chlorofluorocarbons but for all the other assaults we make on the environment—most of these problems would begin to go away.

That would certainly be a good beginning.

Unfortunately, a beginning is all it would be. On that subject, too, there's good news and bad news.

The good news is that a lot of us, thousands, even millions of us worldwide, have already made at least a start on that beginning. We try to recycle our Sunday newspapers and our aluminum beverage cans and as many as possible of the bottles, jugs and jars we lug home from the supermarket. We do our best to use public transportation—or bikes, or our feet—instead of our cars when we can. When possible, we repair our machines and appliances rather than replace them.

The bad news, though, is that that isn't enough to do the job.

In the long run (and the long run is getting shorter every day) governments will have to step in. The changes that are necessary will have to be forced by legislation and treaties, by such measures as prohibition laws or selective taxation.

That's where the hard part begins. That's where we get into such thorny areas as politics and economics—into all the things that would have to be done in order to make it possible to get that kind of legislation passed, when every legislator is under immense pressure to vote the other way, and into what it will mean to all those people who work in industries which will be severely damaged by the steps that must be taken to avert disaster.

We will take a hard look at just how hard that is going to be in this book. For now, though, there's no reason for any of us to put off our own commitment to the things we *can* do as individuals. Until serious government action happens, or even if it never happens, each one of us can do his mite.

Of course, what one person can do to save the environment is no more important, on a global scale, than what one person can do to harm it. Still, there are two real benefits we can claim from our individual actions even if we can accomplish nothing more.

The first is that we will get, and deserve, the respect of our children and thoughtful people of all kinds.

The second is less tangible. It is this: If each one of us uses only the sustainable amounts—even though most of the rest of the world's people go on in their profligate ways—we may not by ourselves be able to forestall disaster. But at least our consciences will be at ease, because we will know that we are not *accomplices* in the destruction of our planet.

And—who knows?—there may even be a third benefit. Maybe we'll set such a good example that others will be inspired to follow.

* * *

All that sort of talk sounds pretty arrogant, doesn't it?

It is arrogant. Face it, that's the way we conservation-minded types are. We are arrogant enough to think that we know what the world needs better than our families, our neighbors and even our governments . . . and on these issues, unfortunately, we do.

Let's go one step farther. Let's admit that when it comes to making decisions on what must be done about all this, we know better than even the scientists who study these matters.

This doesn't mean that the scientists are not to be trusted. On the contrary, we can't get along without their knowledge, as far as it goes.

It's true that scientists are only human beings, and so some of them may be self-serving. A few may be mistaken, or even incompetent, and we know, sadly, that there are a handful who are simply dishonest. But by and large the world's scientific community includes some of the brightest and most able men and women alive, and they devote their lives to the effort to understand what is going on in the physical world.

If they say—as collectively they do say—that, for example, our burning of fossil fuels is leading to serious changes in the world's climate, then we must believe that the information they give us is true. They are the experts.

That's as far as their expertise goes, though. It does not extend to telling us precisely what to *do* about that information.

The scientists are like the expert witnesses before a grand jury. Their only job is to tell the jurors what they know. Sometimes what these witnesses say will be confusing, or unclear, or even contradictory, but that doesn't let the grand jurors off the hook. They still have to sort the evidence out as best they can, and then it is the jurors, not the expert witnesses, who have to decide what action to take.

In this case, we are the grand jury. Some tough decisions have to be made that affect all our lives—decisions like how much we are willing to pay for a cleaner and safer world; what inconveniences we are willing to put up with; how much trouble we are willing to go to to make the planet safe for our grandchildren. The scientists can't make these decisions for us, because they don't know any more about balancing these costs and benefits than we do. Even the government can't make the decisions for us—or won't—because the task is too great, and especially because our legislators and government agencies are under too much pressure from the special interests which cripple them; in fact, we will have to be the ones to force our government to act.

It's up to us, the ordinary man and woman in the streets. We have to make these hard decisions as best we can, because there isn't anybody else.

5

Where We Go from Here

What's going to happen in the next seven chapters is that we will describe the assaults on the natural environment that are presently taking place in the world, and what damaging consequences they will produce.

These assaults aren't all equally important. Some are relatively trivial, some are life-threatening. For convenience we can group them into five different levels of environmental destruction and list them in ascending order, like this:

The first, and least threatening, are what we might call the continuing esthetic or moral losses to the world.

That includes such things as the approaching extinction of many attractive animal species (elephants, butterflies, whales, songbirds and others) and of a far larger number of species of plants; plus the despoiling of natural treasures like forests and riversides. These are things that make the world a nicer place. Most people would agree that it would be a pity to lose them. In a few places they represent serious economic losses to the local people—from tourism in the African game parks, for instance—but few people are likely to die because of their loss.

Second, there is the loss of possible future benefits from sources we haven't discovered yet.

The most obvious example of this lies in the destruction of natural forests, particularly the immense tracts of tropical forests which have never been completely studied. In such places there are many kinds of living things which are being identified only after they have already become extinct. By then it is too late to know whether they have any practical value. We can be statistically sure, from the experience of history, that among them we would have found a fair number of such things as new pharmaceuticals and useful varieties of food plants—if we hadn't allowed them to be wiped out before they were identified. We can be quite certain that we are losing treasures in this way. What we don't know, and can never know, is exactly what those treasures might have done for our health and well-being.

Third, there is the loss of amenities and benign environmental conditions.

Here we include the damage to lakes and waterways through pollution; the destruction of forests from acid rain and uncontrolled lumbering; the increase in smog levels in our cities. These things will certainly cost us a lot in terms of both money and health, but (with some effort) their loss should be survivable for the bulk of the human race, if not for the unfortunate smaller number of people whose supplies of potable water and conditions of personal health are most damaged.

Fourth, there is severe global-scale damage to the environment through the "greenhouse effect" warming.

This is a hairy one. No one can say in precise detail what all the effects of a continuing and significant global warming would be. The best we can do is say that it is highly probable that among them would be climate changes in major food-producing areas that would have the effect of cutting productivity. This could well imply famines, even some quite large-scale ones. Still, at worst the warming does not really threaten the survival of the entire human race—if only because, in the worst possible case, it would be self-limiting: so many people would die as the result of runaway climatic changes that there would not be enough survivors still around to continue the process.

Fifth, there is The Big One: the total extinction of almost all life on Earth.

That's not an easy feat to accomplish. We probably couldn't successfully do it with nuclear weapons, even with an all-out nuclear war. But there is a chance that we might do it yet with spray cans and foam plastic. If we should let the process go far enough to lose the ozone layer completely, we would then have destroyed our only real defense against the high intensity ultraviolet radiation—it is not much of an exaggeration to call these components of sunlight "death rays"—from the Sun. Many of our most indispensable food plants would die and, ultimately, so would we.

Most of this book primarily concerns itself with American problems and what can be done about them. There is a reason for that.

Of course the damage to our environment isn't a purely American problem. It affects every human being alive, from the Eskimos to the inhabitants of Tierra del Fuego, but there are two reasons why we chose to concentrate on the American scene.

First, America is the country which is doing the most damage to the environment.

The United States has only one-twentieth of the world's population, but we generate the most heat-retaining carbon dioxide, manufacture the greatest volume of ozone-destroying chlorofluorocarbons, have the most exhaust-emitting automobiles, generate the most fuel-burning electrical energy, and thus contribute the most to the world's ecological problems.

There's nothing here for Americans to take much pride in. On the other hand, it isn't a confession of unusual sin, either. We don't have to beat our breasts any more than anyone else, because the rest of the world isn't any more virtuous than we are. With a few honorable exceptions—almost all of them in Scandinavia—*every* community and every nation in the world is busily polluting our planet just about as fast as its population density and level of industrialization will let it.

What makes America different is its size, its wealth and the fact that it got started early, and so it has a good, long head start on the rest of the world.

The second reason for concentrating on America is that, of all the countries in the world, the United States is in the best position to do something important about the problems.

Americans are generally pretty well educated. They have excellent access to the media of news and information. They possess the habit of free elections and a well tested political mechanism for running their government . . . and the country *is* rich. For all these reasons, the United States is the nation that can best afford to take the lead.

More than that, America has just shown the world how effective it can be in leading world opinion in the mobilization of international opposition to Saddam Hussein after his invasion of Kuwait. Others protested. America acted. Within hours after the Iraqi troops attacked, the American government, under President George Bush, had begun a whirlwind drive of enlisting support, laying plans, securing the passage of United Nations resolutions and beginning the buildup that led to Saddam Hussein's military defeat. It is possible to question the motives behind this remarkable show of leadership. It is not possible to doubt that it worked.

Regrettably, the same government has shown no interest at all in leadership in environmental actions. In many attempts at international cooperation, the way the United States has used its power has been to try to slow down or even prevent agreement on reforms.

The USA could not save the world all by itself. The problems can't be solved in any one country, however advanced. The solutions require worldwide action. If the USA became saintly overnight it would only postpone, not prevent, the day of reckoning as long as the rest of the world continued on its present course, and even that temporary postponement would be very brief.

But still, it is America that should take the lead—not because it is the guiltiest country, but because it is the country that is best able to do so. Not for altruistic reasons, either—at least, not *just* for altruistic reasons—but because in the long run that is the only way it can save itself.

THE
PROBLEMS

THE
PROBLEMS

6

Warming Up

The world we live in is getting warmer because of what is called the "greenhouse effect."

There isn't really any doubt about this. In August, 1990, the United Nations brought together the 300-plus leading specialists in this area from more than twenty different countries to analyze the situation. They specifically examined the theories which cast doubt on the reality or the importance of the warming.

Now the only real questions left are how serious the consequences of the warming will be, and whether we've already begun to see its first effects. There are still some skeptics, and there is still a lot of debate about specific details. But the greenhouse effect is real.

What the "Greenhouse Effect" Is

What causes this greenhouse warming is the presence of a collection of gases that trap heat in the air, of which the principal members are water vapor, carbon dioxide, methane, nitrous oxide and the synthetic gases called chlorofluorocarbons (or CFCs—the same ones which attack the ozone layer, as we shall see).

There isn't very much of these gases in the atmosphere— altogether only a few hundred parts in a million—but they have a large effect. They store heat. The gases are quite transparent to sunlight, so the Sun's rays pass through them with little effect on their way down to the surface. But when the heat of the Sun is absorbed and re-radiated from the surface of the Earth its frequency changes. The re-radiated energy becomes long-wavelength infrared. The greenhouse gases are affected quite differently by those altered frequencies, soaking them up and absorbing their heat. This causes the gases to become warmer than the air around them. Then they pass on that trapped heat to the rest of the atmosphere, and thus to the whole world we live in.

As one would expect, the more of these gases there are in the atmosphere, the more radiated heat they are able to trap and the warmer the world becomes. (After a certain limit is reached this effect lessens, but not enough to do us much good.) The bad news is that these gases have been increasing for nearly a century and a half, since the start of the Industrial Revolution. The worse news still is that over the last few decades even the rate of that increase has been picking up speed.

What the Greenhouse Gases Are

Before we can learn how to deal with these gases we need to know more about them. Let's begin with the most famous of the greenhouse gases, carbon dioxide.

Carbon dioxide comes from many sources: from forest

fires, from decaying waste, even (though not enough to make a difference) from our own lungs. But the major new source is now the burning of such fossil fuels as oil, coal and natural gas.

What makes the burning of fossil fuels important is that they are, in fact, fossils. The clearest example is coal, which is what remains of trees and other plants which lived millions of years ago. They died and were buried by sediment; then, over the millennia, age and pressure turned them into the coal our miners dig up for us to burn.

If you were to chop down a tree today and burn it for firewood, all of the carbon it contains would become carbon dioxide. If you dug up the coal formed by a tree of the same size and burned it, you would release the same amount of carbon dioxide. What is different about the two processes is a question of time. The carbon in the tree you just cut down was removed from the air only within the last few decades. The carbon in the coal is much older; it has been safely stored away, kept from the air, for millions of years. The carbon in oil and natural gas is equally old, and has been out of circulation equally long. But now we are adding these millions of years of accumulated carbon to the air in a matter of decades. "Fossil fuels" are still being formed—but we are burning them at a rate about a hundred thousand times as fast as they form.

That is why the amount of carbon dioxide in the air continues to grow. Starting in the middle of the last century, we began burning coal in the early steam engines, railroad locomotives and steamships, and the proportion of atmospheric carbon dioxide grew accordingly. There were 270 parts of carbon dioxide to a million parts of air in 1850. By 1957, as the burning of oil and natural gas began to be added to, and even exceed, the burning of coal, that had grown to 315 parts per million—and by the end of the 1980s to more than 350. Currently the carbon dioxide content of the air is increasing at the rate of 3.6% a year, and it may be picking up speed. Every time we burn a gallon of gasoline in our car we add twenty pounds of carbon dioxide to the air. Every time we burn natural gas in our home furnaces, or burn oil or coal in an electrical power generating station, we add more—and we use more and more energy every year.

Of course, when carbon dioxide is produced and enters the air, it does not remain there passively forever. "Gaia" is—which is to say, natural processes are—at work here, too. Some of it is dissolved in the ocean. More important, a great deal is taken up by plants as they grow. Every two-ton tree in the forest has in its lifetime taken nearly seven tons of carbon dioxide out of the atmosphere and transformed it into its own woody substance.

The world's forests could do a great deal to help us deal with the excess carbon dioxide production if we let them. We aren't doing that, however. We're cutting them down—more than forty square miles of Brazilian rain forest disappearing every day, 25,000,000 acres of our own Pacific Northwest woods logged flat every year, similar deforestation going on all over the world—and so day by day the carbon-dioxide heat trap grows stronger.

(As if that weren't bad enough, Freeman Dyson points out there is another worry connected with the burning of fossil fuels. For every three tons of fuel we burn we also burn eight tons of oxygen from the air, and the amount of oxygen in the air worldwide seems to be decreasing by about 13 parts per million every year as a consequence. It would take a long time for that to so deplete the atmosphere's oxygen that we would have trouble breathing. The oxygen dissolved in the world's oceans, however, may be more at risk. According to Dyson, "The Pacific Ocean is already seriously depleted," and he recommends a global project for measuring oxygen concentrations everywhere as a priority measure.)

The Other Greenhouse Gases

Carbon dioxide is not the only culprit in the warmup. Unfortunately for us, some of the other gases are even better at trapping solar heat.

One particular group of these other gases have never before had any effect on the Earth's heat balance. They couldn't, since until now they didn't exist. They don't occur in nature. They are man-made synthetic gases such as the chlorofluorocarbons (which we will discuss in more detail later on, in their capacity as ozone-killers).

These CFCs are useful in one or another kind of industry (refrigeration, spray cans, electronic manufacturing, etc.), and so we have begun to manufacture large quantities of them over the last few decades. The effect of this on the global climate has been that the CFCs now account for 25% of human contributions to the greenhouse effect. That is certainly bad enough, but it's worth remembering that it could have been even worse. If the U.S. hadn't begun to impose some limits to CFC-using spray cans in the 1970s, those chemicals would now be adding more to the annual increase in solar heat retention than even the carbon dioxide itself.

The remaining major greenhouse gases—methane and

nitrous oxide—generally come into the air from natural sources, but some of those sources have been increased drastically because of human activities.

Methane comes largely from bacteria in swamps and similar places, but there is a large component which comes from the digestive tracts of animals. The cattle we raise for beef and milk produce methane as they ruminate, and emit it as the odorous gas that permeates cow barns. About a twentieth of the weight of their fodder is transformed into methane—a hundred million tons of it a year. All animals do the same thing, of course; cows are particularly efficient at it, because of their large size and the nature of their digestive tracts, but even creatures as small as termites do their share. The crepitation of a single termite can't compare to that of a cow, but there are a lot more termites than there are cows.

Of course, termites can't be blamed on human farming; there aren't any termite ranchers. There are rice farmers, though, and rice paddies produce prodigious quantities of methane as the organic matter they contain rots under water. The same airless rot goes on in garbage dumps; the gas you smell coming out of one contains more methane. And a very large contribution of methane comes from coal-mining, particularly from the deep shafts which expose to the atmosphere carbon sources that may have been locked away for millions of years.

In the long run, this methane may turn out to be the hardest of the greenhouse gases to control. It is very effective at trapping heat (molecule for molecule, it is as much as 30 times as good at it as carbon dioxide), and its concentration is growing comparably rapidly, at about the rate of 1% increase each year.

Worst of all there is a great deal of methane around that doesn't affect us now because, like the fossil fuels before we began digging them up, it is at present held harmlessly in storage underneath the surface of the Earth. This "fossil methane" may not stay there. Its source is buried organic matter; about 14% of all the organic carbon in the world is in the sort of frozen sherbet of degraded vegetation mixed with water ice that is called tundra or permafrost. It isn't buried very deep, but as long as the tundra stays cold the carbon will stay where it is. But as the world's climate warms, millions of square miles of tundra (mostly in Siberia and the Hudson Bay–James Bay area of Canada) will warm with it. If the warming reaches a high enough level bacteria will start turning that carbon into methane.

There's a similar trapped methane source at the bottom of the oceans—a lot of organic carbon is stored there, so much that this underwater store contains more carbon than the total of all the coal reserves on the planet. If that undersea deposit warms up enough, the gas will come bubbling up in "methane plumes." In some places it already does. Soviet scientists observed one such plume, a quarter of a mile long, in the Sea of Okhotsk a few years ago.

For that matter, the very soil that our plants grow on is loaded with organic carbon—that is what helps to make it fertile—and if global temperatures rise by as little as another 3 degrees as much as 250 billion tons of that stored carbon from the soil may be released, either as methane or as carbon dioxide, as the global warming proceeds . . . thus accelerating the process still again.

There isn't much to say about that other greenhouse gas, nitrous oxide, simply because much of the needed data about its sources is still a mystery. (What we do know is somewhat surprising; for instance, about a tenth of the nitrous oxide in the atmosphere appears to come from gases produced in the manufacture of nylon stockings and pantyhose.) Similarly, we might as well skip over the greenhouse gases we haven't previously mentioned, such as water vapor (because, although it is a powerful factor in warming, there's not much we can do to keep the sun from evaporating moisture from the sea) and ozone, because we'll have a lot to say about that in a future chapter.

The total effect of them all, though, is formidable. The trace gases are extremely efficient at the greenhouse effect, for which reason methane, nitrous oxide and the CFCs collectively contribute nearly as much to the warming of our planet as carbon dioxide itself. And the quantities of those gases in the atmosphere are rising faster than the carbon dioxide.

How Bad Is a Greenhouse?

Now that we've looked at what the greenhouse gases are and how they work, we might reasonably want to ask if it all really matters.

When you first think about it, living in a greenhouse may not sound like such a bad thing. For those of us who live in what are called the temperate zones, with their often bitterly cold winters, there are times every January when a little global warming seems pretty attractive.

It doesn't work that way. Warming the Earth will not turn Long Island into Tahiti. One of the most damaging effects of the greenhouse warming is likely to be a significant increase in violent weather, followed by drastic and rapid changes in the climate conditions many living things depend on for their survival.

The reason for this is that the atmosphere is basically a heat engine. The more heat energy that the greenhouse gases trap in the atmosphere the more it has at its disposal to transform into kinetic energy—the energy of motion—the energy we see as winds and weather.

That is a simplified statement of a complicated matter; actually, kinetic energy is produced by *differences* in temperature. If the world were to heat up evenly at all points the increase in the movements of the air would be minor. Since that is unlikely, the energy added by the greenhouse effect will express itself in the form of blowing winds and the movement of air masses from place to place.

This means more, and more damaging, violent storms.

We may already be seeing this happening, in such devastating recent hurricanes as Gilbert and Hugo, in the summers of 1988 and 1989. Hurricane Gilbert, which killed 300 and left more than a million homeless in Jamaica and Mexico, was called the most destructive hurricane that ever struck that region—until the next year, when Hurricane Hugo turned out to be even worse.

These Atlantic hurricanes do terrible damage, particularly to the barrier islands and coastlines of the United States from Florida north. Will the damage increase?

It looks that way for two reasons. First, there appears to be a cycle of hurricane violence, with a period of fifteen to twenty-five years. Studies just published by William Gray at Colorado State University showed that, although the total number of hurricanes is fairly steady year after year, the number of the most damaging storms is not. The twenty-three

years from 1947 to 1969 averaged about 8.5 days of very violent Atlantic hurricanes from 1947 to 1969, while in the period
from 1970 to 1987 that dropped by three-quarters, to only
2.1 days per year . . . and in 1988-1989 rose again to 9.4 days
a year.

That suggests that the Atlantic is now entering another
period of a couple of decades of violent storms, which the
global warming can only make worse. All the strength and
destructive power of hurricanes come directly from the
warmth of the oceans where they develop, and one scientist
has predicted that the global warming may produce future
hurricanes as much as 25% stronger than those of the recent past. Some, like Hurricane Hugo, are already tremen-
dously powerful. No one knows just how fast Hugo's winds
blew, because they blew the anemometers right away, but
one instrument registered over 160 miles an hour before it
was lost.

That sort of wind velocity is actually worse even than it
sounds. The force of impact of blowing winds goes up as the
cube of their velocity; thus a 200 mph wind will cause not just
a quarter more destruction than a 160 mph one, but nearly
twice as much.

Not only hurricanes will increase. There are two other
kinds of terrifyingly destructive storms that develop from
contrasts between heat and cold: the tornado and the microburst.

A recent example of the first of these is the killer tornadoes and wind storms that struck Huntsville, Alabama, and
Newburgh, New York, in mid-November, 1989. The weather
conditions at that time were such that a very warm air mass
had covered much of the eastern United States for some days,

setting new records for high temperature readings for that time of year in many cities. Then a very cold air mass came down from Canada and pushed into the warm air. That is a classical meteorological situation, only a bit more severe than usual. The border between two such masses of air is called a "cold front." Meteorologists draw lines to represent such things on their charts—you can see cold fronts marked on the maps of most TV weather programs—because when a mass of cold air meets a mass of warm air on such a front it is their collision which spawns most of the precipitation that occurs, and produces the most severe storms. The greater the contrast in temperatures, the more violent the tornadic storms are likely to be.

Almost all Americans have known what tornadoes are like ever since one of them carried Dorothy off to Oz, but the microburst is a fairly uncommon phenomenon which we are just learning to dread. The microburst is the mirror image of the tornado. Where a tornado is produced by the winds that circle around a spiral of rising warm air, the microburst is the product of the winds around a rapidly descending mass of cold air, which is falling from the top of a thunderhead to the ground below. When the cold air hits the ground as a microburst it splashes out in all directions, producing instant wind gusts of up to 150 miles an hour.

That is what pilots call "wind shear," and it can be deadly. When such a gust of wind comes from directly behind a plane coming in for a landing—even a relatively mild gust, of no more than 30 or 40 miles an hour—it cuts the plane's airspeed drastically in a moment, thus causing the aircraft to lose the lift that kept it airborne. Wind shears of this sort have caused more than 30 United States plane crashes since

1964, including one in New York City which killed 113 people, and they are likely to become more common when weather worsens. (Fortunately for future air travelers it is beginning to be possible to detect these wind shears by radar and warn pilots in time to avoid them—at least part of the time.)

Microbursts can be even more deadly than tornadoes, though happily they are not only much rarer but also much briefer. Tornadoes last for a while. In daylight they can be seen as they approach as funnel-shaped clouds that reach down to the ground and move across the landscape; it is frequently possible to take shelter from them, or even to run or drive out of their way. Microbursts come without warning. In the morning of June 29th, 1990, a microburst hit Streamwood, Illinois, near Chicago. It flattened an industrial park and was gone—in a matter of seconds—while beyond a radius of a few hundred yards everything around was almost untouched.

It is not only increasingly serious storms that the greenhouse effect offers us. Surprisingly, even record cold snaps may be a symptom of the warmup.

On January 20, 1985, the Chicago area set a new all-time low temperature record—27 degrees below zero, Fahrenheit. Pipes froze, cars would not start, shrubs died. Yet that cold air was not any colder than usual. It was just in a different *place* than usual—brought down from the Arctic by the movement of air masses. The same is true of the killing frosts that wiped out the Florida tomato crop in December, 1989, and of other untimely recent freezes in many parts of the world.

In the past couple of years a number of places on the Earth

have been struck by what were called "once in a century" weather events. There was a flooding rain in the Chicago area in August, 1987, that was described as the kind of thing that happened only once in a century—until it happened again, in the same place, just a few weeks later. England, with no experience in its recent history of hurricanes, suffered one in October, 1987, that caused half a billion dollars damage and toppled ancient trees all over the United Kingdom. (In the same year storms in South Africa killed 174, in that country's worst-ever natural disaster.) The 1987 British storm was also called a "once in a century event"—until one almost as bad came again in October, 1989, and a third and even a fourth occurred in January and February, 1990, with gusts up to 155 miles an hour and 46 persons left dead.

Perhaps we now have to redefine "once in a century" to mean "once in the *last* century," for the present century seems to be operating under new meteorological rules.

Although meteorology has come a long way in the past few decades it is still very far from an exact science.

If anything, climatology—the study of overall climates, rather than of the rapidly changing weather systems—is even less exact. To begin with, the very term "climate" is hard to define with scientific precision. The easiest way to describe it is in layman's terms: "Climate" is the kind of weather you expect to have, most of the time. The Iraqi desert, for instance, is said to have a dry climate—that's why it's a desert. Still, as we saw in the Persian Gulf air-war early in 1991, it can have heavy enough rain and cloud cover on occasion to interfere with aircraft missions, though such conditions are not usually expected.

The climate of any area is determined primarily by two factors: Its distance from the Equator, which affects its average temperature, and the direction of its prevailing winds. The effect of the winds, in turn, is affected by where they come from (oceans or dry land, colder areas or warmer ones) and by what intervenes in their paths before they reach the target area. If the winds pass over a mountainous region, much of the moisture they may contain is drained out of them as they are lifted and cooled. Furthermore, all marine air masses are affected by currents in the ocean, which transport warmer or colder water from its place of origin to regions as much as several thousand miles away—as the Gulf Stream, for instance, transports some Caribbean warmth as far north as Iceland and the British Isles.

That sounds very complicated, and it is. That is why it is impossible to describe in detail just what effect the global warming will have on every region of the Earth; there are too many factors, they are not all well understood, and there are even indications that, even if all the factors were known, long-range weather forecasts would still be impossible because the processes appear to be "chaotic"—a technical term meaning that factors too small to identify may wind up as having very large long-term effects. The usual way this is expressed is to say that even the flutter of a butterfly's wings in Mexico may result as a tornado in Missouri. Of course, it also may not. That's what chaos means.

Still, it is possible to model the probable consequences of the greenhouse effect on local climates with at least fair confidence. Such computer models have been made, and some of these effects are quite unpleasant to anticipate.

It is almost certain, for instance, that the predicted changes in weather patterns would seriously damage some of the most important farm areas in the world. The American midwest— the "breadbasket" of America, and indeed of much of the world—would be particularly hard hit. In that marvelously productive farmland protracted droughts are expected to occur because of the greenhouse warming, as precipitation patterns change and strong winds dry out the soil. (It's worth remembering that some of that part of America was called by its first explorers "the Great American Desert.")

This forecast of climate change does not predict that rainfall will cease all through the American midwest. Rain will certainly continue to fall in the area, and indeed almost everywhere else on Earth. As a global average, annual rainfall totals may even increase, for as the Earth warms, evaporation from the oceans will increase. Sooner or later all that added moisture will have to fall out as precipitation; it has no other place to go.

But the precipitation will not necessarily occur where and when we human beings find it useful. One scenario for the American corn belt indicates there actually will be more rain coming down in an average year—but it won't be as useful for farmers, since the bulk of the increase will come in the winter months, while rainfall in the summer, when the farm belt's crops need it, will actually be less than normal. The rain will certainly fall; it just won't fall in that particular place. The 1988 midwest drought offers an example of what is likely to happen: in that expensive case a strong high-pressure area pushed the jet stream north, so that its accompanying rainstorms fell over Canada. As a result in that summer, Illinois farms received less rain than the semi-desert of New Mexico. (Nor will the winter snows lock up moisture to release in the spring thaw in the same way they do now. The warmer winters will melt the snows early, allowing them to run off uselessly before crops can benefit from them.) Another American scenario, this one farther west, suggests that the Colorado River will lose approximately 29% of its water if there is a two degree rise in the temperature of its basin, with serious consequences for the already water-starved areas that depend on it for their water, such as Las Vegas and much of Southern California.

Some of the projected changes in precipitation patterns can give us the worst of both worlds: drought and flooding at the same time, when occasional cloudbursts flood rivers over their banks while leaving reservoirs still empty. We have already seen something of that sort in parts of southern California, and especially in England in the early winter of 1991, when torrential rains doubled the width of the Severn River and people living along its banks were still forbidden to use tapwater to wash their cars.

Then there is the problem of ocean currents. These are basically driven by air temperatures. As the Earth warms, the courses of the currents of the sea may be altered.

The general pattern of water circulation, particularly in the northern hemisphere's Atlantic Ocean, is fairly well understood. As cold water sinks to the bottom of the sea around the Arctic Ocean, it travels slowly southward along the seabottom toward the Equator. Once it reaches the equatorial tropics it is warmed and accordingly rises to the surface, after which it slowly moves northward, in the form of the ocean surface currents mariners have long learned to map, making a complete circuit as the water returns toward the Pole.

That's where one serious problem may arise. One scenario suggests that the warming at the North Pole (which is expected to increase in warmth somewhat more than the global average) will provide less and less of that cold bottom water that drives the whole system as it flows south along the sea bottom. Without this constant thrust of cold northern water, the Gulf Stream—the principal vehicle for the surface return flow in the Atlantic—will lose power. It may slow down, stop or even reverse its direction of flow.

That could be quite serious. Since England depends for its gentle climate on the warmth it receives from the northern end of the Stream, it is bad news for Londoners; without the Gulf Stream, the usually mild British winters will become more like Labrador's.

There is another grave problem associated with the oceans. If the warming is allowed to continue, the sea levels will rise throughout the world.

As a matter of observed fact, the sea is already rising in many places around the Earth. Two-year-old highways in the Seychelle Islands are already unusable because the sea has covered them; coastal farm areas in Asia and South America now need dikes to keep the sea out. In particular, the city of Venice has suffered severe flood problems. Many other reasons that have nothing to do with global warming, such as tides and pollution, are involved in Venice's case. Nevertheless, one significant factor that contributes to the occasional need to cross St. Marks Square in boats is the rise that has already been measured—actually, it is no more than two or three inches so far—in the average level of the Adriatic Sea. Some of that appears definitely to be due to warming; it is known that the deep water in all of the western Mediterranean

Sea, which had been quite constant in temperature from the beginning of the twentieth century until the middle of it, then warmed by a small amount—about a quarter of a degree Fahrenheit—between 1959 and 1989.

How much of a global sea-level rise can we expect, over what period of time?

Here again, the answer is far from clear, since so many factors are involved. We do know some of them. We know, for instance, that some of the worldwide rise will be due to the melting of trapped ice—for instance, such huge accumulations as the Greenland ice cap and glaciers all around the world. Most of those are already retreating, in fact. The white caps that top the Alps are glaciers, and they have been melting steadily since the beginning of the Industrial Revolution a century and a half ago; already, half of the 1850 level of Alpine ice has melted and run off to the sea.

Most worries about melting ice do not center around the northern ice masses, however, for 90% of all the ice in the world is in the Antarctic.

The Antarctic ice is not likely to melt to any large extent, at least for the next century or so. True, there are some disturbing indications. The whole Antarctic region is known to have warmed by about 1 degree Fahrenheit since accurate records began to be kept in the 1957 International Geophysical year. The ice caves at McMurdo Sound, one of the favorite tourist sights for Americans stationed there, are now off limits because these unusually warm temperatures have weakened the ice and made them unsafe to enter. And some great icebergs, larger than ordinary, have been observed to calve off the Antarctic ice shelves in recent years—one more than two miles long, containing a hundred million tons of ice, off the Erebus Glacier on March 1st, 1990; large-scale calving from the Ross and Filchner-Ronne Ice Shelves has increased as well; while the relatively small Wordie Ice Shelf, which floats in the sea off the Antarctic Peninsula, has been actually breaking up since 1958.

But that Antarctic ice cap, which is as much as two miles thick, has survived climatic changes for ten million years. In spite of some local apparent thawing, there is no clear observational indication of widespread, large-scale melting in the Antarctic ice cap.

The generally accepted theoretical understanding of Antarctic ice is somewhat reassuring—at least, in that it predicts that if large-scale melting did occur it would not happen quickly, but could only take place over a thousand years or more because ice is such a poor conductor of heat; whatever happened to the surface, the deep-down ice wouldn't even "know" it—at least, wouldn't be affected by it—for some time. (On the other hand, the generally accepted theory also predicts that the whole West Antarctic Ice Shelf should have melted away long ago. It hasn't; so the theory is not completely trustworthy.) In any case, Antarctic ice will take longer to melt simply because it is generally a good deal colder than northern ice; the Alpine glaciers are often only a degree or two colder than the freezing point of water, while the Antarctic ice in many places is as much as eighty degrees colder than that.

There are even some theoretical reasons to think that, at least in the early stages, a global warming might actually *increase* the amount of water stored in the form of ice, and thus kept out of the world ocean, in the Antarctic. Warmer oceans would produce more evaporation; more water vapor in the air would produce more precipitation worldwide; in Ant-arctica the extra precipitation would fall as snow and stay locked there in the form of ice.

However, there too the theory contains some loopholes. The breakup of the Wordie Ice Shelf, for instance, suggests a troubling scenario. Calving and breaking up of the floating ice shelves doesn't affect the sea level; it is only where the ice rests firmly on land that its melting can raise the sea level. The ice shelves have another important function, though. Just by the inertia of their presence they slow down the gradual creep of the land-bound glaciers toward the sea. If they break away, the land ice will slide down to become sea ice more rapidly, and that *will* add volume to the sea. So, once again, we have another set of interactions whose con-sequences are hard to predict.

It is worth remembering for future reference, though, that if the Antarctic ice cap did melt, its melt-water would add nearly 300 feet to the world's sea level—the height of a 30-story building. (Melting the Arctic ice would not have the same effect because most of that northern ice is already floating in the sea.)

Unfortunately, we don't know enough yet about the dy-namics of our icecaps to predict precisely what will happen. Even more unfortunately, some elements of the necessary scientific data that we could have had by now have been de-nied to us for reasons of military security. Scientists would give a great deal to be told, for instance, just how the thick-ness of the Arctic ice is changing. That information has already been obtained, for Soviet and American nuclear-powered submarines have been playing tag with each other under the Arctic ice for nearly forty years, and every one of them has routinely measured the ice levels overhead with echo sounders as they went. All that information is stored away in the records of their respective naval authorities . . . but it is classified. Neither the Soviet nor the American government has been willing to release this useful information for civilian use, on the grounds that it would reveal too much about what their submarines do.

Melting ice is not the only factor in raising the sea level, however. There are many smaller ones, including the fact that a good deal of water is stored on and under the land, much of it in underground sources like the Ogalalla aquifer; the more farmers pump out of their wells, the higher the sea level will rise as that water flows to the ocean.

But that is a relatively tiny factor. The most important one is quite different; it is thermal expansion.

As every high-school science student knows, liquid water does not expand or contract very much with changes in temperature. But it does do so a *little*, and there is so much water in the seas that even that small amount of thermal expansion will raise the sea levels as they warm. In fact, it has already begun to do so.

How far will the seas rise, and when?

Here, as with so many other questions about the environment, we don't know the precise answers, for the familiar old reasons: because those questions did not seem important enough to worry most people until fairly recently, no one was willing to fund the necessary research. Scientists do not doubt that the process is going on, but they can't yet give the answers they want to such highly interested parties as, for instance, the owners of Florida beachfront property or the builders of Dutch dikes.

In the long run, water added anywhere raises the sea level everywhere. All the oceans are interconnected, and water seeks its own level worldwide. The problems in getting specific arise from the fact that the process is complicated by tides, currents and "oscillations" in which masses of water surge back and forth across whole oceans. The process of reaching equilibrium also takes time. When a glacier calves in an Alaskan fjord or an iceberg cracks off an Antarctic ice shelf the sea levels all over the world will certainly rise a bit as a result—but not uniformly, and not all at the same rate. The same with thermal expansion. The currents of the sea go in many directions, driven and directed by surface winds, by geographical features (such as the slope of the sea bottom), and by differences in the temperature and salt content of the water, which affect its density. Any change at one place affects all the others, but the change is registered only as the currents carry it.

The best current mathematical models make it clear that some places will be affected more than others. One study, for instance, shows that thermal expansion will be greatest in the deep water of the North Atlantic, while near Antarctica the level of the Ross Sea may actually decline slightly.

With all those reservations, some things are quite clear. One is that even a small rise would be enough to endanger low-lying shore areas. The United Nations Intergovernmental Panel on Climate Change gave its best estimate of that, estimating that a three-foot rise in sea levels (one possibility for sometime in the next century) would submerge nearly a quarter of a million miles of the world's coastlines. This, they said, would "seriously threaten low-lying urban areas, flood productive land, contaminate fresh water supplies and destroy many coastal wetlands." That is only approximate, of course. As with all these projections, it might turn out that the situation did not become that bad. But no one should take too much comfort from that, because there is an equal probability that it could in fact turn out to be worse.

In specific cases, the effects of a warmup may be concentrated—or reduced—in some quite unexpected ways, as the general warming starts triggering other mechanisms.

Imagine, for instance, that the seas rise on the coasts of Bangladesh and Guyana, which are particularly at risk. On average, an increase of an inch in sea level means an encroachment of sea on land of six feet or more. This figure is for an ideal average coast. Flat lands such as these two will be flooded much farther inland.

Suppose that therefore in these two places many square miles of what was once dry land will become shallow bays. Now some of the secondary effects begin. There is more sea surface exposed to the heat of the sun, from which it follows that there is a bit more evaporation every day. This has its own consequences. There is a negative effect on water height, as the loss of water tends to lower the local sea levels slightly again. But there is also a positive effect. That evaporated water enters the air as water vapor, which itself is a greenhouse gas. Thus it increases the local temperature, causing more expansion. At the same time, some of that evaporated water turns into clouds, which screen out some of the sunlight in their shadow and thus tend to reduce the evaporation again.

All these things are known qualitatively, as scientists say—that is, it is known that they *do* happen—but to put the right numbers on them, to make them quantitative, in order to show just how *much* their combined effects will alter the general picture, is still very difficult. Nor do these relatively simple factors take into account the most catastrophic threat to coastal areas, which is from a probable increase in storms, and thus of storm flooding.

"Sea level," after all, is only an *average* figure. It does not allow for exceptionally high tides, and it does not put a limit on how high the water can rise when storms occur.

In hurricanes and typhoons (which are the same extreme weather phenomenon under different names), the low barometric pressure at the heart of the storm actually pulls the water level up. As meteorologists describe it, a fall of 1 millibar in atmospheric pressure raises the sea level by almost half an inch. In a recent Shanghai typhoon, the resulting rise in sea level coming up onto the land, greatly aided by the strong storm winds, amounted to 22 *feet*. In the probable worse storms of the global warming, such storm surges can be confidently expected to be even larger.

So the coastlines of the planet, where a third of the world's cropland is located and where more than a billion human beings live, will be at considerable risk—if not from permanent flooding, at least from devastating storm surges. Especially around some shores with steep bluffs, where a sea-level rise means there will be deeper water near the land, which means the waves will be higher and more damaging.

Even miles from the coast people who live along tidal rivers will share the risk: that is London's situation, which has already forced its people to build great mechanical barriers just downstream of the city. Inhabitants of cities which draw their drinking water from partially tidal rivers will experience saltwater pollution of their water supplies as the brine-laden tidal bores reach farther and farther upstream from the sea, while those whose water supplies come from underground aquifers will find the salt seeping into those as well. At the same time, the wetlands that rim most shores will be drowned out, with very damaging consequences to their ability (already compromised by drainage and building) to provide nurseries for sea life.

The uncertainties we have pointed out are real, but they should not obscure the general conclusion: if the Earth warms, as it seems clear it must go on doing over some fairly short period of time, the seas *will* continue to rise, and that will endanger many hundreds of millions of human beings. A large number of them will in fact die of it.

In the United States, the state of Florida is at particular risk. Its geography makes that certain. Much of the state, and most of the most heavily built sections of it, are coastal. The whole state is at low elevations above sea level. Increasing storms will certainly flood the coastal areas more often. Ultimately many parts of Florida may be permanently drowned under shallow seas.

The Mississippi delta is also highly vulnerable; one scientist predicts that by 2030 the Gulf of Mexico coastline will be on the southern edge of New Orleans. New York City may need Dutch-style dikes to protect at least lower Manhattan; Boston may need the same; while the barrier islands of the

Atlantic coast, from Florida to Massachusetts, many of which can barely survive the sea's onslaughts now, are likely to be drowned out by the sea early on.

In other parts of the world the situation is even more serious. The country of Bangladesh, which lies in the unstable deltas of three major (and unpredictable) rivers, already experiences disastrous and repeated river flooding. In 1987, 40% of Bangladesh's area was flooded, the worst such experience on record—until 1988, when the flooded area reached 62%. Worse still are the cyclones and storm surges from the Bay of Bengal. 300,000 of Bangladesh's people died there in 1970 in such a seawater flooding, tens of thousands more in 1985. 110 million people live in Bangladesh, 80 million of them in threatened areas; uncontrolled sea-level rising can obliterate their homeland just about entirely. Nor is it likely that a country as poor as Bangladesh can find the money to build sea walls as Holland has done to keep the water out; the cost of coastal protection is estimated at as much as ten times the cash value of the lands it would protect. (One serious proposal for Bangladesh is that it forget about trying to protect its coasts and instead spend what money it can afford on bulldozing tall earthen mounds scattered over the lowlying areas, so that people and their livestock will at least have a place to retreat to when the land is flooded.)

Indonesia is almost as endangered as Bangladesh. An island nation, it has one of the world's longest coastlines, with as much as 40% of its total area at risk (and, ominously, Indonesian government resettlement programs are currently relocating more people into those coastal areas every year). On Taiwan, the Vilan plain is one of the country's most productive farm areas and even now needs the protection of dikes to keep the sea out.

No inhabited continent is without its endangered areas. In Africa, Gambia, and in South America, the countries of Surinam and Guyana, are at grave risk already. On the eastern coast of Guyana, in the district called Berbice, an artificial sea wall like Taiwan's is all that protects some three thousand square miles of sugar plantations and other farms, the richest agricultural land in the country; Guyana has already had to spend almost a third of its annual construction budget on building and maintaining these dikes, but it is running out of money and the dikes have already begun to crumble. Farther south along the continent's Atlantic coast, most of the hilly city of Rio de Janeiro should stay dry, but not its famous Copacabana and Ipanema beach areas. Neither will some suburbs of Sydney, Australia; or parts of such low-lying European cities as St. Petersburg and Hamburg.

The direst threat of all is to such island nations as the Maldives. The highest point anywhere in the Maldives is only seven feet above sea level. In the event of rising seas and major storms its inhabitants will be in even worse straits than the people of Bangladesh, for there will simply be no dry land for them to run to.

The coral islands of the Pacific are equally threatened, though in other ways. Most of them are volcanic in origin, like Hawaii and Tahiti; but in most cases those stark volcanic cones have been added to by thousands of years of coral growth.

Coral is particularly vulnerable to increasing warmth. If the temperature of the water it grows in rises too high— which may be only a degree or two above normal—the coral dies, "bleaches," turns white as the living organisms (they are called zooxanthellae) that contribute to it expire. And once the coral has begun to bleach, it can no longer renew itself and sooner or later the action of the seas will wash it away.

This has been observed to happen in recent years, particularly in the Caribbean. In 1987 and 1989 (1988 was spared because Hurricane Hugo cooled things off), coral bleached in massive quantities around the island of Jamaica. The water temperature had reached 86 degrees Fahrenheit—not much above the normal summertime range of 82 to 84 degrees, but enough to kill.

Closer to home, the Florida Keys Reef, the only real coral reef in the mainland USA, has been showing signs of bleaching in recent years—a situation made worse by increasing algal growth (encouraged by sewage from the Keys' growing communities) which chokes the coral.

What puts coral islands particularly at risk from the global warmup is that they are seriously threatened by other processes. It is not only warm water that threatens coral. It dies when it is choked by sediment coming off the nearby land (as in many tropical islands, where the hillsides have been denuded of trees and soil erosion is acute); or when it is damaged by nuclear explosions, as in France's bomb-testing range in Polynesia; or when it is crushed by ocean vessels running aground, as has happened twice to the beautiful corals in Florida's underwater park; or when it is attacked by predators like the Crown of Thorns starfish that has damaged so much of Australia's Great Barrier Reef. When the dead coral can no longer repair itself it disintegrates, and then the rest of the island is no longer protected by its sheltering reefs.

* * *

So far we've looked at two of the major problems associated with the global warmup: the changes in weather, especially as they affect our agriculture, and the rise in sea levels. One other big problem remains.

That is what is likely to happen to the world's remaining forests—"devastation" may be the best word to describe it—from the consequences of the warmup itself.

We are not speaking now of human lumbering activities or man-made forest fires, but of the things that the changes in climate will bring about as they occur. Nor is it only forests that will be affected. Every kind of plant will be affected, simply because they are plants.

One of the obvious differences between plants and animals is that plants can't move. Where their seeds fall is where they will grow, and where they will die. If things go wrong in that place they have no way of escaping to another.

As the Earth warms, the zones of climate will move away from the equator and toward the poles; in the United States, they will move north. Missouri will become more like Louisiana, Iowa more like Oklahoma, Saskatchewan more like South Dakota.

This will have damaging results for the surviving forests, since trees, like every other living thing, have evolved to survive most successfully in the climatic zones they are in. This puts them among the most vulnerable of living things to climate change. Animals—most of them, anyway—can migrate to a more comfortable place when the weather changes . . . as long as there is a place to go. (That won't be true of every animal. Many endangered species, such as African elephants and the Chinese panda, now survive only in limited preserves. They are locked into their present territories because they are surrounded by farms, towns and other places where they cannot survive. In a warming they will have no place to go and they will almost certainly become extinct.)

Trees can't migrate, and so changes of climate mean death for them.

This is not a new thing in the history of the planet. The fossil records show how often such a change has wiped out a whole species of vegetation. But the fossil record is not an accurate picture of what will be happening to the world's trees (already, one would think, more than adequately threatened by acid rain, agricultural clearing and uncontrolled lumbering) as we artificially warm the Earth. Natural climatic changes take a long time. Although the individual tree can't move, over the time involved in that kind of slow change the forests the trees grow in (and other "climax" vegetation) can creep along to follow the changing climate. If they had enough centuries to make the transition, hardwoods would simply slowly replace balsam firs in northern Minnesota, broad-leaved plants would take over from grasses, grasslands would replace sugar maples and oaks in Michigan. There would still be some extinctions. But some sort of vegetation would survive.

Not in this case. The necessary centuries of time will not be available for them, and changes like these can't occur overnight. Each individual tree is rooted to the spot where it grows, and so the forests they grow in can move only generation by generation. In a warming period the forests need the years and decades that are required for the mature trees on the cooler, northern edge of the zone to spread the seeds that will grow new saplings a little farther north, while the warmer, southern edge dies away and is replaced by vegetation suitable to a warmer climate.

In a man-made, accelerated global warming the forests won't have that kind of time to adjust. They will simply die.

There are, of course, other effects associated with poleward movement of temperature zones that will affect plant life. Our farmers are not bound to the timetable of forests. Farmers can plant their crops wherever they like; they don't depend on the slow natural spreading of species. But the crops they plant must be the ones suitable for a particular locality. The crops which now thrive in Kansas may well find the temperatures and rain patterns they need farther north in Alberta. Whether they will thrive there is another question: northern soils are generally thinner and less fertile soils; but in any case the patterns farmers are used to, and know how to deal with, will no longer apply.

Farmers will have another problem, since they will also have to face insect pests which will follow the changing climate to their present croplands. As the winters grow warmer, insects which now are killed with the first frost will be able to survive overwinter and multiply in the spring. Even the familiar local pests are likely to become more active (and thus more destructive) with extra days of warmth in the year.

Not only crops will be endangered. Insects which carry diseases that affect human beings will follow those that prey on plants: malaria, perhaps even yellow fever and other scourges, will no longer be confined to the subtropics. (Human health will be affected in other ways, too, some of them rather unexpected. There are non-infective ailments which are encouraged by warm weather; according to one medical projection, with hotter temperatures more Americans will suffer from such diseases as kidney stones.)

Of course, in the long run—in the *very* long run—we wouldn't be worrying about kidney stones, floods or dying forests. We would have far worse things to worry about. Given just the right conditions—enough burning to turn all the world's fossil fuels into smoke, enough warming to release all that stored carbon in tundras and sea waters and turn it into carbon dioxide—there is a real possibility that the global warming would become a runaway, feeding on itself, and never stopping until there was no longer enough free carbon and oxygen left in the world to react.

That is only a theoretical fear. But if you would like a scary look at what a real, runaway carbon-dioxide situation is like you can do no better than to open an astronomy textbook and consider the planet Venus.

Venus is the ultimate example of a carbon-dioxide greenhouse that has gone too far. Its atmosphere is almost pure carbon dioxide, and its surface temperature, at nearly 900 degrees Fahrenheit, is enough to melt lead.

Some might say that that's not too surprising, since Venus's orbit lies much closer to the heat-radiating Sun than our own. That's true enough, but it isn't the basic reason for Venus's deadly heat. The other inner planet, Mercury (which is essentially without an atmosphere of any kind), is far closer to the Sun than Venus. Nevertheless, Venus is hotter than Mercury. It is the hottest planet in our solar system.

Its dense carbon dioxide blanket is what has made it so.

Will that happen to the Earth?

No, of course not. It could not reach that condition in the same way as Venus—through an increase in carbon dioxide alone—because the chemistries of the planet are different. It is not likely to have that sort of runaway from the synthetic greenhouse gases we manufacture, either. In spite of the very best we human beings can do, and are doing, to wreck the planet we live on, it is not ever going to reach a Venus-like state.

We shouldn't take much personal comfort from this, however. The principal reason we can be quite sure that it can't happen here is only that the process would take a long time, and well before our planet reached that state we human beings—the motive force driving our world in that direction—would all be long dead.

Perhaps it's time to take a recess from this long and depressing catalogue of greenhouse effects; enough is enough.

What we have said doesn't cover everything that can be said on the subject. Perhaps it doesn't even answer all the questions that you may have in your mind—questions that might begin with something like, "Is this all *real*?"

Perhaps it's time for a little reality-testing in general. We'll try to do that in the next chapter.

7

Question Period

While we're taking a breath let's try to answer some very reasonable questions about what we've said already. There are three good ones:

"Won't natural checks and balances take care of keeping the world's climate within a normal range, as they always have? I mean, what about *Gaia*?"

"If Gaia can't help us, is there anything we can do about it on our own?"

And—the most important question of all—"Tell the truth: is all this scary stuff about global warming *real*?"

There are short answers to all these questions. In order, the answers are: "No," "Yes, but not easily or cheaply," and, "Unfortunately, yes." But the long answers are more useful.

Can natural checks and balances save us from the effects of the man-made greenhouse global warming?

No, they can't. They are certainly doing their best, in the face of greater challenges, due to our uncontrolled mining and burning of fossil fuels, than they have ever had to face before, but their best is not good enough.

As we've already seen, there are many natural processes

which operate to remove carbon dioxide from the air (they are technically called "sinks"), such as coal formation and the deep-freeze of dead plant material in the tundras. There are a great many of these natural sinks; but most of them are much too weak to resist the large-scale changes in the atmosphere.

Only two sinks, really, are big enough to make a dent in the problem. The first of them is the world aggregate of living plants, which suck carbon dioxide out of the air to turn it into vegetation. The other is the world's bodies of water, mostly the oceans, which dissolve large quantities of the carbon dioxide out of the atmosphere in much the same way that a cup of coffee dissolves a spoonful of sugar.

The world's living plants are our most valuable sink. When a plant takes in a molecule of carbon dioxide from the air, it uses sunlight to break that molecule down into its component elements, carbon and oxygen; the process is called photosynthesis. The plant uses the carbon that results to build into its own structure; that's how it grows. The plant has no use for oxygen, however, so it exhales the oxygen back to the air—that is how, billions of years ago when plants began to appear, the Earth came to have an oxygen atmosphere for us to breathe in the first place.

Plant photosynthesis is a very satisfactory process for human needs—indeed, you could call it indispensable, not just for us but for almost all life on Earth. But we are killing the plants that do the job. We cut down our trees in vast numbers for such trivial reasons as to splinter them into disposable chopsticks for export to the Orient, as well as to use them for paper and lumber or simply to get them out of the way so we can build houses and plant farms.

The habit of destroying woodlands did not begin with our

generation. Scientists tell us that the pristine world—the world that existed before the human race became numerous enough, and aggressive enough, to have much of an effect on it—contained almost twice as many forested acres as now remain; it was human beings who destroyed the missing half. But those ancient people who burned or cut down all those trees did not have our chain saws and bulldozers, so they were nowhere nearly as good at it as we are.

It isn't only trees that we are removing from the cycle. We take our toll of everything that is green. We destroy more of that indispensable plant life every time we drain a swamp, or bulkhead a shoreland, or construct a highway, or start a new suburban housing development, or lay a concrete parking lot for a shopping mall. And we do all this on so vast a scale that the vegetative sink is no longer large enough to handle the task of keeping the carbon-dioxide level in check.

As to the sink in the oceans: Water is very good at dissolving carbon dioxide—that's what makes our fizzy drinks like beer and ginger ale possible. Over the ages, the seas have dissolved enormous quantities of the gas, so that now there is far more carbon dioxide dissolved in the global ocean than is free in the atmosphere.

But there is a limit to how rapidly the oceans can take up excess carbon dioxide, and that limit depends on the temperature. If the seas should warm enough, more carbon dioxide will bubble out of solution than there is being dissolved—in just the same way that when you leave ginger ale out overnight it goes flat faster than the same beverage kept in the refrigerator.

In that case, the race that we call the carbon-dioxide exchange between the atmosphere and the oceans will go the wrong way. It will make the situation worse instead of better.

(Nor is the situation any better for the other greenhouse gases. The sinks for methane are similar to those for carbon dioxide, and equally threatened. For the synthetic gases like the chlorofluorocarbons there are no natural sinks at all. None ever arose, since those gases did not exist until we began to manufacture them.)

So Gaia alone can't do the job for us any more. It was primarily through these two carbon-dioxide sinks that she managed to keep the global climate so stable over so many millions of years, but now we've crippled her.

Which brings us to the second question: Is there anything we can do to restore Gaia to health—say, by inventing some technological fix?

That is the question with the "Yes, but" answer. There are a lot of things we might be able to *try* in the attempt to remove surplus greenhouse gases from the air, but no obvious good ones. Some would work too slowly to do us much good, some might not work at all, some might actually make the situation worse.

For instance, we can start planting new trees on the barren lands where the forests have been cut down.

That's the most obvious (and lowest-tech) of all the current proposals. It would probably help considerably, in the long run, but it is an enormous job. To replace the cut-over regions would mean planting the greatest forest in the world, covering an area about the size of Australia.

Reforestation on that scale may well be worth undertaking for many other reasons—for instance, in order to control soil erosion—but it can't solve our present problem. The atmospheric benefits of replanting logged-out areas come too slowly to help us now. It takes anywhere from forty years to several centuries for a newly planted sapling to grow big enough to match the carbon-holding capacity of a mature tree. We don't have that much time to spare.

There have been more technologically "sophisticated" proposals by the dozen. Some are grotesquely unrealistic, though they come from respected institutions: for instance, a Brookhaven National Laboratory project to catch all the carbon dioxide that comes from the world's smokestacks and pump it through vast pipes to the bottom of the sea, where it can dissolve and remain out of the air, at least for a while. (Think of the cost of such a program! Think of the extra carbon dioxide that would be produced by the power plants that ran the pumps that forced all that gas against the pressure of the deep sea.) Two Japanese scientists have a similar idea; they also want to catch all the stack emissions, but then their idea is to pump them through great tanks holding a soup of algae, letting the algae do what plants always do. (Expense again; plus the fact that suitable algae strains would have to be genetically engineered; plus the problem of how you then dispose of the slurry of dead algae that results.) Some are simply inadequate to the task, like the Environmental Protection Agency's March, 1989, suggestions of cutting down on cement production and finding alternative methods of producing rice, meat and milk. (Again worth doing for other reasons, but not for dealing with the surplus carbon dioxide; the EPA plan would affect only the other greenhouse gas, methane, and only a small fraction of that.) Some might help to ameliorate the carbon-dioxide problem—like the EPA's other suggestion, of replacing fossil-fuel power plants with nuclear reactors—but create serious problems of another kind. (Such problems as the expense of constructing the nuclear plants; the long time delay involved in building such plants, which typically run to a decade or more; the dangers of nuclear plants—such as Chernobyl-style accidents—and the equally worrisome but generally overlooked unsolved problems that afflict all such plants, such as the total lack of disposal facilities for their radioactive wastes.)

There are more speculative proposals, too. They represent wishes more than solid, realizable plans, but they are worth a look.

For instance: If natural trees grow too slowly, how about letting the molecular biologists build us some new kind of tree or shrub—one as good at surviving and growing as any weed—that will come to maturity very rapidly, and, what's more, will do it almost anywhere in the world without special care or irrigation, so that we could perhaps spray seeds out of airplanes over the bare Asian hills, or even over the empty Sahara? That's not quite impossible. Conceivably such new tree species could some day be bred. But we don't have them now, and no one presently knows how to start creating them; so that is a wish rather than a plan.

All right, then: How about dispensing with vegetation entirely and going right to inorganic chemistry?

We know that a lot of carbon dioxide does get taken out of the air by chemical means and turned into rocks like limestone. That doesn't happen fast enough naturally to be useful to our present needs, but perhaps science can find a way to speed it up. Can't we just spray some magic fairy dust into the air (let's not call it that; let's call the stuff by the more sober-sounding chemical name of "catalyst")—a high-tech catalyst, then, which would make the process happen quickly, creating great masses of carbon-containing rock which we could then bulldoze underground—or simply leave lying harmlessly around, or even use for building materials?

No. We can't. We don't know how; but even if we did we probably wouldn't dare. Like the sorcerer's apprentice, we would be tempting fate if we started things we might not be able to stop.

If we did have the capacity to speed up some such natural process we would risk it going beyond our control. Then it might keep on removing carbon dioxide past the point we intended it for—perhaps until it got down to that 200 parts per million level that means another Ice Age—or even beyond that, perhaps to the point where there is no atmospheric carbon dioxide at all. That would make our planet almost as cold as Mars. Then we would all die.

So we can't put back the lost trees in time, and we can't expect the inventors to give us a gadget that will make the process stop.

There's only one thing left. We're going to have to do it the hard way . . . if we are to have any hope of doing it at all. That is, we're going to have to cut down on the amount of fossil fuel we burn.

That doesn't mean we have to forswear the use of fossil fuel entirely. That wouldn't be sensible—there are applications to which fossil fuels are vastly better suited than any imaginable alternative—but fortunately it isn't necessary. What we have to do is *reduce* our burning of coal, oil and

natural gas to the point of our Steady-state Allowable Perturbation, or SAP. So let's try to estimate just how much
man-made interference we can inflict upon the environment
in this area without making things worse.

That means we have to put numbers into the equation.
Fortunately, in the race between our production of greenhouse gases and their removal from the atmosphere, we do
know some of the necessary numbers. We know, for instance,
that the human race now turns out some 50 billion tons of
carbon dioxide a year from the burning of fossil fuels. And
we know that that's too much.

It's doubtful that there is any way at all to stop the process
of global warming entirely, whatever we do. The process has
picked up too much momentum, and stopping or reversing
it may be simply beyond our capacities. But perhaps we can
do the next best thing by slowing it down. If we can do that
adequately, then we can at least allow the world time to adjust to the coming climate changes.

We have a number for that. A pair of scientists, one from
the Lawrence Berkeley Laboratories in California, the other
a climatologist from West Germany, have prepared an estimate of what a tolerable rate of global warming should be.
Their conclusion: no more than an increase of a twentieth of
a degree every ten years. Any rate of global warming higher
than that would result in unacceptable damage to forests and
crops.

Even that is very rapid, as natural global change goes. (The
warming after the last Ice Age took place at an average of
about a *thirtieth* of a degree every ten years.) And it is certainly far less than our present practices would produce.
According to a United Nations panel which studied the

question in April, 1990, we're likely to experience an increase of about a whole degree every ten years between now and the year 2030. That's twenty times the proposed maximum level.

If we translate that one-twentieth of a degree level into what is produced by carbon dioxide production, holding our temperature rise to safe limits means that we have to make quite drastic cuts in our burning of fossil fuels. Our present discharge of 50 billion tons a year of carbon dioxide would have to come down to about 6 billion tons a year—about as much as we now produce every 44 days.

So, as a first approximation, anyway, that can be our SAP for carbon-dioxide production. It's formidable. It means that for every gallon of oil we now burn, we have to restrict ourselves to a little less than a pint. (And that doesn't even take into account what has to be done about the other greenhouse gases—particularly methane and the CFCs.)

It isn't quite as bad as it sounds, though. We don't have to reduce our *energy* consumption to the same degree—as we will see, we can start taking advantage of non-polluting energy sources to make up much of the slack; and we can save a good deal of the rest by using our energy more efficiently.

Still, reducing the rate of warming will certainly be a difficult task, and it is likely to be an expensive one as well.

Just how much this will cost us, in public and private funds, is almost as hard to predict as the fine-screen detail of what the climate changes will mean in your own neighborhood. Some studies are fairly scary. One such modeling of the probable cost of doing what is necessary to achieve this sort of steady state puts the price of change, for the United

States alone, at anywhere from $800 billion to more than $3 trillion—and several times that when the changes for the rest of the world are taken into account.

But, in the event, it isn't going to be that bad. Those high cost-estimates show only the loss side of the ledger; there are gains as well.

The money we will save from, for example, more efficient use of energy will go a long way toward meeting those costs. Some estimates even suggest that a conversion to a steady-state energy economy will actually mean a *profit* for the world, in terms of cheaper bills forever after.

Later on we'll go into the details of costs, benefits and strategies. For now let's just conclude that it is possible to stave off at least the worst effects of the greenhouse disaster. The ways of doing it are known. It is only the dislocations and difficulties involved in making the transition to a steady-state world that are hard to face.

So now we come to the most important question of all:

Is all this *real*?

As we said earlier, the short answer to this is "yes." But let's look at the evidence for this statement, so you can make up your own mind.

Start with the fact that most of the world's experts agree on that answer. Since August, 1990, we have a reliable scientific answer to that question in the words of the Intergovernmental Panel on Climate Change. It began its report by saying: "The scientific community agrees on the climate changes that could occur if the present buildup of greenhouse gases continues into the next century."

That may not settle the question for everyone, since some people would still like to dismiss these predictions of catastrophic warming as "mere theory."

Well, the predictions *are* theoretical, of course. They have to be. That's the nature of the beast, since the worst of them haven't happened yet.

But some theories need to be taken seriously. When you were a child and your mother told you not to touch a hot stove, because it would burn you, she was making a theoretical prediction—based on what she knew about the nature of hot stoves and your fingers. If you then tested it out, you discovered very soon that her theory was quite "robust," as the scientists say—meaning you found yourself yelling and sucking your finger.

We obviously don't want to test out these environmental theories in quite the same way—if we checked them out by letting them happen the pain would be a lot worse. Nevertheless, there are ways to try to decide how "robust" they are, and some of them we can use for our own decision-making.

We've already stipulated that, like any prediction, these can't be exact in detail. That can't be helped. Almost any theoretical prediction, including these, is subject to an "error bar"—that is, an estimate of how much higher or lower the actual value of the thing predicted might be. (The error bar in your mother's prediction about the hot stove lay in her uncertainty about whether you would wet the tip of your finger and touch it lightly, or press your whole hand against it.) Since there are serious gaps in our understanding of all these environmental processes, their error bars are considerable. And some scientists point out that it is well within these margins for error that things might turn out to be far less damaging than we have described here.

This is true—theoretically. But it is based on a very unlikely implication. For that to be so it would have to happen that almost *every* error went in the *same* direction, and that that direction was toward the least harmful outcome.

But that isn't the way to bet it. It is just as likely that they will all go in the *wrong* direction, making things far *worse* than we have described . . . and most likely of all that whatever errors there are will go in random directions, more or less canceling each other out, so that these predictions will be reasonably close to accurate.

The next thing we might want to do is to ask ourselves how reliable the theory is. If we are to take these forecasts as a basis for action, we should try to reassure ourselves that they are at least theoretically valid.

Fortunately there is a way to help us make that kind of decision. We can judge them just as any other scientific theory is judged.

Traditionally, the proper test of any scientific theory rests on the predictions that can be made from it; that's the heart of what is called "the scientific method." If, on examination, the predictions turn out to be right, then the theory can be taken to be an at least approximately reliable description of reality—good enough to deserve to be acted on, anyway.

So what can be predicted from the theories we have discussed?

One prediction would be a rise in sea levels; and we have seen that that is happening in many cases. Another would be an increase in violent weather—also observed. A third would be a detectable melting of fossil ice—the ice that has been around for a long, long time—and we've seen that there is evidence of that happening, too.

Most important of all, we would find that the temperatures all around the world are generally getting higher—not much, of course, because these things happen slowly, but all the same enough to be observable.

That sort of evidence is harder to find than you might think, because the simple and unpredictable day-to-day variations can be scores or even hundreds of times larger than the predicted annual increase. But some indications are there, all the same.

You can't go by what happens in a single locality, of course. Still, it's interesting that in England, for example, August, 1990, produced some of the highest temperatures the country ever recorded—as much as 100 degrees, an outlandish reading in England—so high that a Caribbean sea turtle was spotted in the English Channel and at Mountfichet Castle, in Essex, the waxwork guard who stood on top of the tower melted away. (As a tour guide said ruefully, "All that's left is a Norman helmet and two glass eyes in a pool of wax.")

The global mean temperature is more meaningful. It goes in the same direction. Of the seven hottest years on record, six have come since 1980. 1990 was the hottest of all.

This was predicted exactly by a greenhouse scientist named James Hansen, who offered to bet anyone who cared to take his wager that 1990 would set a new global record for temperature. Many scientists disagreed, but only one took the bet. That one was Hugh Elsaesser of the Lawrence Livermore Laboratories; and in January, 1991, Elsaesser ceremoniously wrote Hansen a check for $100, because the evidence was in and it showed that Hansen was right.

Does all that *prove* the theories?

Not quite. We're not entirely out of the "error bars" yet. The actual increase in the average temperature worldwide is still tiny. But what we can say for sure is that these things are precisely what we would expect to find if the theories were true, and that we would be very surprised to find them if the theories were false.

On that basis, it seems less useful to debate the question of whether all this is really real than to think about what we should *do* about it.

We will surely do that later in this book, but first we have other things to discuss. The global warming is not the only great environmental problem we face. Before we get into the question of what we can do about global warming (and of what we can do about the problems those actions will produce), first we need to examine what those other problems are like, since they are largely interrelated and—thank heaven!—the measures we need to take to deal with one problem will often help solve another one as well.

Even our assaults on the air we breathe don't end with warming it up; in the next chapters we will look at some of the other equally ominous processes involved.

8

Poisoning the Air

The same fossil fuels that contribute to the greenhouse effect when we burn them don't stop there. They do us damage in other ways as well. They pollute the air we breathe. They destroy forests with acid rain. They dirty our homes, and our lungs, with soot. They corrode the stone in ancient monuments and kill the fish in ancient lakes. They shorten people's lives and damage their health. They sicken children, and in some parts of the world they kill them.

The fossil fuels don't do the job alone; they have help from industrial poisons and agricultural ones and from many other sources, but almost all the sources have one familiar thing in common: As with just about every one of our environmental ailments, they cannot be blamed on God's will—or on Gaia's. They are entirely the product of what we do to ourselves.

One of the problems in talking about air pollution is that it sounds like yesterday's problem. Most of us can hardly remember a time when there wasn't concern about all the man-made poisons in the air, and demands that something be done about them. Most of us even think that something *has* been done already.

Indeed something has—slowly and grudgingly, to be sure; but nevertheless it's true that governments have taken at least some faltering first steps to abate some of them. Through the federal Clean Air Act and thousands of state and local ordinances American oil companies have been forbidden to add lead to their gasoline, saving many children from sickness and mental retardation; compulsory catalytic converters have somewhat reduced the poisons from automobile exhaust pipes and factories; and power plants have been compelled to scrub out some of the contaminants from their smokestacks.

But the problem isn't solved, or anywhere near it. What has happened is that we now generate new pollution faster than we clean up the old, even in America—and anyway it is only the United States and a very few other countries that have done even that much. Most of the world is still lagging behind in even these basic measures . . . and here again the problem is global. We all breathe the same air. From Hudson Bay seals to the penguins of Antarctica, the bodies of every living thing on Earth—including every human being—are carrying measurable amounts of airborne man-made poisons.

What has happened to the problem of air pollution is only what happens to most of the world's environmental problems. The problems don't get solved. They simply get pushed aside, because they are swamped with unexpected newer and even worse ones; and so yesterday's problem is still very much today's.

To understand what the problem of air pollution is all about, and to have some hope of finding ways of dealing with it, we need to know what it is like.

To begin with, there are six major classes of pollutants in the air we breathe. They are—not necessarily in order of importance, because which substance does the most harm varies from place to place—ozone, acid rain, carbon monoxide, soot, toxic chemicals and radioactive gases and particles.

Ozone is worth discussing first, if only because we know that its role in our lives is so confusing: Ozone is definitely a poison, but it is also indispensable to our survival. In the next chapter, we will see how ozone is a great friend to man—when the ozone is located high above the Earth in the stratosphere, where it is our only protection against the lethal forms of ultraviolet radiation from the Sun.

At the surface level, however, that very same ozone is a dangerous pollutant.

That shouldn't surprise us. Almost every pollutant—in fact, almost every kind of waste—is actually a useful substance that happens to be in the wrong place. In the case of ozone, it would be very nice if we could somehow trap all that damaging surface ozone and pipe it up to the stratosphere, where it not only does no harm but is urgently needed to keep us from the killing ultraviolet rays. We can't do that. Nor will the surface ozone oblige us by floating up to the stratosphere by itself. It doesn't survive long enough to do that. What makes ozone so destructive is that it is one of the most violently reactive chemicals known. For that reason, it is bound to have a chemical reaction with something else—very often by burning the lungs or tissues of some living thing, to that organism's serious harm—long before it can rise far above its source on the surface.

The ozone we find near the surface of the Earth comes largely from the cars we drive. The mixture of organic compounds in the gasoline engine's exhaust react with each other, with the help of sunlight, to produce the ozone. Besides being poisonous to breathe, ozone has a bad effect on farms; according to the Environmental Protective Agency the ozone in America's air cuts crop yields by about 12%, a cash loss to farmers and customers of more than $2 billion a year.

Acid rain is almost as damaging as ozone to living things, and there is more of it. Acid rain falls from clouds just as every other rain does. But, like all rain, it dissolves some of the chemical compounds in the air into itself as it falls. What makes the rain thus produced acid is the presence of large quantities of sulfur and nitrogen compounds in the air. The sulfur comes mostly from the burning of oil and coal. Some fossil fuels are higher in sulfur than others, but almost all of them release some sulfur when burned. The nitrogen comes out of the atmosphere itself—four-fifths of our air is nitrogen. Nitrogen does no harm to us while it is in the air in its natural state, but when any kind of fuel is burned at high temperatures, particularly in car engines, some of the nitrogen in the air is inadvertently "burned" along with the gasoline; it comes out as nitrogen oxides, which then mix with water in the air to turn into nitric acid.

Carbon monoxide is a poison gas which is produced when any carbon fuel does not burn completely. Carbon monoxide is the gas that kills people who commit suicide by starting their car engines in a closed garage, which is a measure of how lethal it is when the concentrations are high. It damages health, largely by interfering with the blood's ability to transport oxygen to the brain and other organs, even when there is not enough of it to kill outright. Nearly all the carbon monoxide in the air of our cities comes from car exhausts.

Soot is the most visible pollutant from combustion; it is the ugly black stuff that we see coming out of factory chimneys and truck exhausts and call "smoke." Soot is made up of little particles of solid matter which have not burned completely. The particles are small enough so that they float in the air for a fairly long period before they precipitate out on your windowsill—or your lungs.

It is soot which caused the famous zero-visibility pea-soup fogs of London in the old days—and in many other cities, as well. Because soot is highly visible, it was one of the first pollutants attacked by early environmentalists—not always with immediate success. When, years ago, the city of Boston finally imposed strict limits on sooty chimney exhausts, the air was so bad that at first the laws could not be enforced. Inspectors could not see what chimneys the soot was coming from because the smog was too thick; they couldn't identify which chimneytops to report.

London, on the other hand, did finally end its traditional sooty fogs by the simple measure of banning all coal fireplaces. Now the Londoners' fireplaces have no real fires in them at all. The fires have been replaced by electric grates.

Unfortunately, most of the power stations that generated the electricity still burned fossil fuel. That meant only a small improvement in London's air, so, to protect their immediate areas, the power plants tried another tactic. They built very tall smokestacks so that the particulate pollution rose high into the air and only fell to earth as soot hundreds of miles away. Like most technological fixes, that one didn't really fix the problem, it only removed it to a different place. In the final analysis, all London had done was to export its smog, in the form of acid rain, to the lakes and forests of Scandinavia . . . which did not appreciate the gift.

Toxic chemicals come out of trace compounds in fossil fuels, to a larger extent from industrial processes and, probably most of all, from the burning of waste in incinerators.

How dangerous these chemicals are depends on what the waste contains. Years ago most urban trash was pretty well limited to paper, food remnants and other organic materials which did relatively little harm when burned. That situation has changed for the worse. Now even household waste is full of plastics and other man-made substances, and most city waste is also loaded with thrown-away oil, discarded tires, industrial chemicals and other combustibles that produce serious poisons when burned.

Those are the major everyday air pollutants. There is another variety which, fortunately, does not appear in any great quantity very often but can be terribly damaging when it does:

Radionuclides. In some ways the radioactive pollutants are the worst of all. They not only do harm and even kill, they *go on* killing for very long periods of time.

While chemical pollutants are unpleasant enough, most of them are short-lived; if we stopped producing them today, the world's air would be nearly clean again in a matter of weeks. (Of course, we aren't likely to do that.) Some of the radioactive pollutants, on the other hand, survive for tens of thousands of years, and they can go right on causing cancers and birth defects for all of that time.

When we speak of airborne radioactivity, the first word that comes to mind is "Chernobyl." That nuclear power station accident in April, 1986, sent a radioactive cloud over most of Europe—and indeed ultimately over much of the world—and the full effects of that exposure are still being discovered. There are many other sources, however. The testing of nuclear weapons produced about as much radioactive fallout as Chernobyl, years ago, and is still occasionally adding small amounts to the world's air.

Chernobyl was not the world's first large-scale nuclear disaster. It was not even the Soviet Union's, though that fact was kept secret until quite recently. In September, 1957, at the nuclear weapons plant at Kyshtym, in the southern Ural Mountains, spent processing fluid was kept in a large tank, and the tank blew up. The explosion was chemical—some reaction among the countless chemicals in the tank. The contents were highly radioactive. A witch's brew of radioactive pollution poured into the air, much like Chernobyl's, contaminating five thousand square miles of territory. Hundreds died—the exact number has never been made known. Ten thousand other people had to be evacuated from the area; there would certainly have been more, but the Kyshtym area was not populous to begin with; that was why it was chosen for the secret plant. Some old villages were simply stricken from Soviet maps, without explanation, and the whole matter was hushed up—until an émigré Soviet scientist, Zhores Medvedev, pieced the story together from obscure papers in Soviet journals, and in 1990 it was finally confirmed by the Soviet authorities.

Countless other sources of radioactive pollution—from weapons-manufacturing plants to nuclear power stations—also add their mite, and more accidents are inevitable.

The worst effects of radioactive fallout are on living things; they cause birth defects, they make crops inedible and they kill. The effects of Chernobyl alone are likely to cause tens or hundreds of thousands of birth defects and cancer deaths that will continue to crop up for decades to come. Some villages near the affected area report a quarter of their livestock births are deformed . . . and so, tragically, are many children.

Since we human beings pour all these poisons in the air, it seems reasonable that we ought to be able to find ways to take them out again.

To some extent, we can. Technological fixes do already exist for at least a partial solution to some of these problems. The toxic chemicals from burning waste can be limited by careful control of the oxygen flow and temperature in the incinerators, or better still by keeping their sources out of the waste in the first place. Better designed automobile engines and catalytic converters can reduce the quantity of pollutants from car exhausts. Scrubbers on the chimneys of coal or oil-burning factories and power plants can remove the worst of the sulfur compounds before they get into the air. (Then you have to deal with the separate problem of finding a way of disposing of the tarry toxic mess that has accumulated in the scrubbers.) Alternatively, the utilities can choose to burn only low-sulfur fuels, though such fuels cost more and would have the disadvantage of putting a lot of soft-coal and high-sulfur workers out of a job.

At best these measures can only reduce the damage to the air, however. They can't clean the air up completely. Moreover, they're expensive, for which reason the most useful clean-air measures are bitterly opposed by the oil companies, the manufacturers, the utility corporations, the car dealers—and a good many of ourselves.

Even a partial reduction in air pollution is definitely worth having, even at a high cost. To convince ourselves of this all we need to do is to look at the parts of the world that have let pollution run unchecked, for example the countries of Eastern Europe. Perhaps we should start with the story of a place called Nikel, in the Soviet Arctic.

When rich deposits of nickel ore were found on the Kola peninsula, just across the border from Norway, the Russians built mines and smelters and created a town. They named it after its reason for existence—Nikel, in Russian spelling—and for years it was one of the world's principal sources of high-grade nickel metal. Ultimately the mines played out. The smelters remained; and for the last few decades they have been as busy as ever, though the ore they refine now has to be shipped in from other sources.

They are also as polluting as ever. Each year Nikel's smelters produce 140,000 tons of the metal. At the same time they produce half a million tons of sulfur dioxide, which exits the plant through their immense stacks and comes down all around the Kola peninsula, and parts farther away, as acid rain. (The entire next-door nation of Norway collectively emits only half that much.) Around Nikel hundreds of square miles of forest are simply dead. There are no living fish in the rivers. Even grass, when sparse blades of it try to come up in the spring, turns brown and dies.

The people of Nikel are almost as badly off. No one lives there, at least not for very long. Nikel's volunteer workers from other parts of the Soviet Union rarely stay more than ten years—nine out of ten of them, Nikel's doctors say, already with permanent lung damage, in spite of the fact that during their working hours they breathe filtered air through a mask and tube.

The pollution doesn't stay on the Soviet side of the border. It is so bad in the Scandinavian countries that from time to time bands of enraged Norwegians storm across the border in protest demonstrations, demanding Nikel clean itself up. It can't. In the present state of the Soviet economy there's simply no money there, so the troubled neighbors—Sweden and Finland as well as Norway—have chipped in to raise a fund of $50 million to get the cleanup started.

Nikel's environmental situation is atrocious. It isn't alone in Eastern Europe, though. In Cracow, Poland, tourists are advised not to remain in the city for more than three days because of dangerous air pollution. The residents are far worse off. Curiously, the rate of deaths from cancer in Cracow is slightly lower than in the rest of Poland . . . encouraging, one would think, until one hears the doctors explain that it is because Cracow's people don't live long enough to suffer from some of the diseases of aging. In Copsa Mica, Romania, the IMMN chemical works specializes in heavy metals, and emits 30,000 tons of metal-laden soot each year. Copsa Mica's workers are healthy when hired, but a year later they have up to 800 times the permissible level of lead in their blood, and almost three-quarters of them suffer from lead-induced anemia. In Czechoslovakia, as many as 10% of infants born in some areas, such as northern Bohemia, are affected by birth defects, and the sunlight throughout the area is described as "bleak and gray." In the little Czechoslovakian town of Teplice children are kept indoors for as much as a month at a time; in the six worst weeks of the year the school and all its children are moved to another community with cleaner air.

For more than seventy years, the countries of Eastern Europe had no foreign exchange to buy low-sulfur fuels. What they did have was plenty of local brown coal deposits, heavily contaminated with sulfur and other chemicals, there for the digging. Dig they did. They scraped away the soil in vast open-pit mines that will remain a scar on the landscape for geological amounts of time and trucked the coal to their power plants, factories and homes to be burned. The result is that, from Poland to Yugoslavia, the air stinks of sulfur fumes.

Some of these Eastern European countries did make token efforts to clean the air. Long before the unification of Germany, the East Germans passed laws which were very much like the American ones, requiring scrubbers on many industrial smokestacks. The laws were simply ignored, however, for the scrubbers needed to comply with them didn't exist. In all of the former East Germany there was only one factory capable of manufacturing that sort of equipment, and the government hunger for foreign currency was so great that the factory was compelled to sell most of its output abroad.

The same sort of situation obtained in Poland, Bulgaria, Czechoslovakia, Hungary, Romania and Yugoslavia—and indeed in the Soviet Union itself—with the result that Eastern Europe now contains the most heavily polluted real estate in the world. The consequences to human health alone are appalling. In some areas human life expectancy is as much as seven years less than even in comparatively cleaner places nearby. The gross national product of Poland is diminished by 10% from pollution of various kinds—about half of that from workers' sickness and consequent absenteeism. There are areas of the former East Germany where the air pollution is still so bad that doctors order children to be sent away until they reach the age of ten, and where some clinics can't be adequately staffed because doctors refuse to work there—it is too discouraging for them to try to treat the illnesses that only clean air could prevent. Nine out of ten of Leipzig's citizens have illnesses related to air pollution. In Budapest, Hungary, hospitals have installed "inhalitoriums"—boxes the size of telephone booths, where patients wait in line for fifteen minutes of breathing cleaned air; one Hungarian death in each seven is caused by pollution.

Nor are human beings the only sufferers. Nothing alive escapes the blight. Farm animals sicken, too, and even vegetation is damaged or killed; in parts of Czechoslovakia and Poland as many as three-quarters of the trees are damaged or dead from acid rain, and the rest of the region is nearly as bad.

Will this change with the end of Soviet domination of the area?

Each country promises to try its best to clean up; indeed, the Greens in each of those countries were among the leaders in the freedom revolutions. But their problems are nearly insurmountable; they do not have the money for the job. Even if the money came from somewhere, what would they do for power and goods while newer, cleaner plants were being built? And what would they use for transportation?

If Eastern European factories and power plants are obsolete and dirty, their cars are even worse. East Germany provided most of the automobiles for the whole area, and the East German "people's" cars, the Warburgs and the Trabants, are about the most polluting in the world. They operate on two-cycle engines, like an American power mower, with both gasoline and lubricating oil mixed together. The exhaust is a foul-smelling black cloud of pollution. Catalytic converters are unknown. . . .

And that could have been America, if we had not begun to take at least a few first steps.

And if we don't go on to go a good deal farther, it might be America yet.

For we Americans do our full share of despoiling the environment with our cars. The cars are cleaner than Trabis, pound for pound. But they are also much bigger—with bigger engines, requiring more fuel to run them—and there are also a lot more of them.

After the Arab oil embargo of the 1970s, laws were passed to require more fuel-efficient (and thus less polluting) cars. When cheap oil came back the laws were systematically weakened. Now more and more Americans are switching to heavier, gas-guzzling vehicles—to larger cars, with more fuel-burning appliances loaded onto them, and also to vans, station wagons, Jeeps, off-the-road vehicles and even small trucks. Catalytic converters and mechanical improvements may keep down the proportion of pollutants that come from each gallon of gas—but the bigger vehicles burn more gallons, so the total emissions increase.

Worse than that, the number of cars hasn't stayed still; now there are more of them than ever on the roads. Which means not only more vehicles burning gas but also more heavy traffic and thus more time spent in traffic jams—which means more emissions. Gasoline wasted with the motor idling in traffic jams cuts the average miles-per-gallon efficiency of the American car by 15%. That efficiency is nothing to boast about anyway; due to the Reagan-Bush easing of constraints on auto manufacturers, it actually dropped a little at the end of the 1980s.

Even when the cars are not running, their wastes continue to choke the air. The gasoline which is spilled each time the tank is filled, and the leakage out of the tank itself, don't disappear. They evaporate into the air, where they are just as polluting as the exhaust itself. While the junked tires—

Ah, the junked automobile tires. They are a *remarkable* pollution source. No one has any good idea of what to do with the things. There was a time when discarded automobile tires were routinely retreaded or even ground up to make other rubber articles. The advent of the steel-belted radial made that more difficult (you can't grind steel), and supplies of cheap rubber made it uneconomic. The tires can't even be buried very well, for their shape makes them work their way back to the surface after every rain. So they just sit there, in immense lots holding tens of thousands of worn-out tires.

Then, from time to time, they burn.

There are few sights as obviously polluting as a dozen acres of abandoned tires on fire. There was a particularly nasty one near the town of Hagersville, south of Toronto, Canada, in the winter of 1990; 14 million tires, stacked twenty-five feet high and covering an area equal to three city blocks, all went up in flames at once. 1700 people were advised to leave their homes while the fire was out of control and the smoke reached for many miles; yet it was not really unusual. In all tire fires the smoke is dense, choking and, from the chemicals in the tires, highly poisonous. The things burn at a temperature of as much as 2500 degrees, which makes it hard for firefighters to get close enough to try to put the fires out, and the heat flashes the hose water into steam before it can do its work. The burning tires pollute in every way possible. As they burn they exude toxic oils which seep into the ground, poisoning nearby water supplies; but it is into the air that they pour the bulk of their choking black smoke—to be carried by the winds to make people gag and gasp in cities miles away.

That doesn't exhaust the list of automobile-generated pollution. The cars' discarded battery acids poison trash dumps. When the cars are finally junked their air conditioners leak CFCs to climb toward their destiny in destroying ozone, while the wornout cars themselves are often abandoned to litter city streets; all in all, it's fair to say that if Henry Ford's brilliant approach to engineering had been aimed at creating a source of pollution, rather than just a vehicle to carry people about, it would have produced about the same design. The motor car is almost *all* pollution.

So the poisons in the air spread—not just to the choking cities, but even to places where one would think the air would be pristine forever. Trees in California's Yosemite National Park, one of the most beautiful and unspoiled wilderness preserves in the world, are noticeably weakened from the effects of ozone and other air pollutants borne in by the wind from the rapidly developing and smog-generating San Joaquin Valley. Nearby parks like Sequoia and Kings Canyon are in even worse shape; in them, hikers

may experience asthmatic attacks, and 10% to 20% of their
evergreen trees are dead. On the East Coast visibility in the
beautiful Shenandoah National Park has been cut in half
since World War II; while around most large Eastern airports
it has dropped from an average of 90 miles to fifteen or less.
Travelers flying into Honolulu—in the middle of the greatest
ocean on Earth—or into Denver—in the heart of the majes-
tic and unspoiled Rocky Mountains—are dismayed to see
the gray-brown clouds of smog hanging over those cities.

Los Angeles, of course, has become a byword for pollution. Every American city suffers to some extent (86% of city
carbon monoxide comes from car exhausts), but Los Angeles is indeed something special. There in 1988, of the 366
days of that year 176 exceeded the legal limits for ozone concentration, and there are many days each year in that city
when children can't be allowed to use their playgrounds but
must be kept indoors—just like the unfortunate school
children of Eastern Europe.

The rest of the "free world" is not much better off than
the United States; "freedom" obviously includes the freedom
to pollute. The very Alps are polluted; 15,000 miles of ski
slopes and 30 million skiers a year are trampling the famed
eidelweiss almost to extinction in some places, while their
cars produce so much pollution that, in the summer of 1990,
tourists were warned to breathe lightly and avoid excessive
exertion. Indeed, the Swiss Forest Institute estimated in late
1989 that half the Alpine trees were sick from acid rain and
man-made dust, and a third of that number either dying or
already dead. The trees in many other parts of western
Europe are equally affected—some 53% of those in West
Germany alone—as they are in New England and eastern

Canada as well, and there are lakes and rivers in all those which have become biologically sterile from airborne pollution. Marble structures which have stood intact for thousands of years have been eaten away by the acid in the air in this century; even roads and bridges suffer damage once blamed on the salt used for snow removal, but now known to be caused at least in part by sulfates coming out of the air. Even the Arctic is not immune; since 1950 air pollution has so increased there that a "pervasive haze" covers almost a tenth of the area, while in our cities architects seldom call for white facing on their skyscrapers any more; the buildings will not stay white for more than a few months, and in fact the stone itself erodes away. Any middle-aged urban dweller who is unsure of just how bad the city airs have become may ask himself just how long ago it was that he last saw the Milky Way in the nighttime sky.

But it is what we can't see that may be the greatest long-term danger to our health.

The deadliest poisons are invisible to us. No one yet knows how many cancers will be caused by the cloud of radioactive gases that the exploding Chernobyl nuclear power plant spewed out in the spring of 1986; estimates range into the hundreds of thousands from this single accident. Nor is there any accurate measure of how many will die as the result of poisons from the incineration of plastics and industrial wastes. Among the chemicals that are produced in this way— and that we then breathe in—are the notorious PCBs and, still worse, the dioxins. These are so deadly that scientists have been unable to find a level, down to parts per *trillion* (think of it as a single drop in a large swimming pool), at which they are *not* dangerous. And even airborne lead is still a threat to the brains and bodies of young children in some places. We thought when we outlawed lead additives in gasoline that we had, at least, that problem solved. In fact, the level of lead in America's air has dropped by nine-tenths since 1970, but serious amounts remain.

Some of them come from quite unexpected sources. One British report, for instance, tells us that crematoria send as much as twenty pounds of lead each into the air each year. The source? No one is absolutely sure; the best theory is that it comes from the burning of the mercury-amalgam fillings in the teeth of the deceased.

Is there anything we can do to make "fresh air" fresh again?

Of course there is, but, again, not easily, for there is really only one way. As Carl Sagan once said, "You put something into the air and it stays for a long time. We have no way to flush it out." So our only hope for clean air is to stop polluting it in the first place.

It is impossible to get rid of the pollution in any other way. There are, after all, only two traditional ways of getting rid of any kind of offensive waste. One way is to bury the wastes out of sight, as we do with our own bodies after we die. The other is to dilute the wastes until we don't notice them any more, like the smoke from burning leaves in the fall or the sewage poured overboard from an ocean-going ship.

Neither will do us much good here. Burial doesn't work for gases. (It doesn't work very well for solid wastes any more, either, as we will see when we come to them.) And neither does dilution any longer—not in a world of more than five billion human beings, all of them producing as much air pollution as they can. There are now too many of us, and we all share the same air. The exhaust that comes out of our own cars and the emissions that come from our own chimneys will blow away from our immediate neighborhood, all right, but where will we find the clean air to blow in and replace them?

So we must limit the input, which brings us to the familiar question: Can we work out a Steady-state Allowable Perturbation—a "SAP"—that will tell us how much air pollution we can afford to produce?

Again the answer is yes, but with difficulties. The difficulty is that we need at least half a dozen SAPs, one for each of the major kinds of pollutants.

For the worst offenders, such as airborne radioactive materials and such highly poisonous chemicals as dioxins, the SAP should be as close to zero as we can get it. Any step less radical than that will not prevent sickness and death. The best it can do is merely to decrease the number of people affected; the dead ones will be just as dead.

For particulate matter—soot and fly-ash—the SAP could well be nearly zero as well, but for a different reason. Particulate matter is the easiest of all pollutants to prevent. Scrubbers and filters—and proper design of all fuel-burning systems—can eliminate it almost completely. In the case of particulate matter the rule should be: If you can see it, prohibit it.

For the remaining major pollutants—ozone and the chemicals that cause acid rain—the SAP can be slightly more generous. There is no doubt that their present levels are far too high. But they don't have to come down to zero, either. For them, what scientists call "threshold levels"—levels below which there is no discernible effect—can be established.

The virtue of most of the airborne chemical pollutants—their *only* virtue—lies in the fact that they are so active chemically that they will use themselves up in reactions with whatever they encounter quite quickly. For that reason it is possible to calculate each SAP on largely economic grounds—how much are we willing to spend to prevent what fraction of asthma attacks and destroyed trees?

But the really good news about air pollution comes in two packages:

First, *it is curable* without the sort of major changes in our life-style that dealing with the greenhouse warmup involves.

Second, if we cut down our production of greenhouse gases by phasing out as much as possible of the burning of fossil fuels, as it appears we have to do anyway, we will automatically reduce a good deal of the airborne pollution, since the two evils come from much the same sources.

There is one other major problem with our air—not that part of it which we breathe down here on the surface of the Earth, but that layer of the atmosphere, high up in the sky, that protects us from the most damaging radiation from the Sun. That requires a chapter of its own, and we'll look at it next.

9

Sunburn

Question: Why do we need the ozone layer?

Answer: Because it is the only thing that saves us from more sunburns, and worse ones, than we have ever known.

You know what a normal sunburn is like. Your exposed skin turns red. If the burn is bad enough, your skin becomes hot and painful. Then blisters form and break, and the surface layers of the skin, killed by the sunburn, eventually peel away to expose new skin growing underneath it.

What kills the cells on the surface of the skin is a particular range of frequencies in the light from the Sun called "ultraviolet rays."

The Sun produces these ultraviolet rays in large quantities, but they don't all reach the surface of the Earth. We are shielded from the worst of them by the presence of a small amount of the gas ozone in the upper levels of our atmosphere. That is the famous "ozone layer" which is now being eaten away by such man-made chemicals as the chlorofluorocarbons, or CFCs.

In order to see what all this means for us—and especially in order to do anything about it—we will want to know more about what we mean by CFCs, about ozone and, for that matter, about ultraviolet radiation, so let's start by looking at each of them.

The word "ultraviolet" sounds as though it ought to be a color. That's reasonable enough, because it is. The only difference between ultraviolet and any other color is that ultraviolet happens to be a color of sunlight that our eyes are not physically able to see.

Like all the colors we experience in nature, ultraviolet is not pure. It is a mix of a number of related wavelengths—just as "green," for instance, may be a bit bluish, or a bit yellowish, without ceasing to be green.

The particular shades of ultraviolet that do the most harm to living organisms (including us) are the ones that are described by scientists as lying in the wavelengths from about 240 to 320 nanometers. Of that lot, the longer wavelengths, the ones above 290 nanometers, are not severely harmful, if taken within reason. They are the ones that tan our skin when we go to the beach—though if we have too much of even those they will give us sunburn, wrinkles or even skin cancers. This commonly experienced 290–320 nanometer range is called the "biologically active" ultraviolet, and it is sometimes referred to as "ultraviolet B."

The radiation at the shorter end of the ultraviolet spectrum, below 290 nanometers, is something else. We don't ordinarily get exposed to those wavelengths. That's a good thing for us, since that shorter-wavelength radiation is much more dangerous. In fact, that range from 240 to 290 nanometers is actually lethal. Even the relatively benign ultraviolet-B destroys living cells when you get too much of it; that is how it happens that children get the pleasure of peeling away strips of dead skin. This shorter-wavelength ultraviolet does more. It destroys the very proteins and DNA molecules the cells are made of.

This lethal variety of ultraviolet is called ultraviolet-C. Human beings haven't had to worry much about it before now because it doesn't naturally penetrate as far down through the air as where we live—at least, it hasn't succeeded in doing that up to the present time. But the shield that keeps it from reaching us is the ozone layer, which is no more than a tenuous scattering of ozone gas, high above our heads. If this ozone layer should disappear, or be substantially thinned, the ultraviolet-C from the Sun would easily penetrate the rest of the atmosphere. Then it would be able to reach us down on the surface of the Earth. Then, if too much of it got through, we would die.

Ozone is a colorless gas whose composition is very simple. It is made up entirely of atoms of the element oxygen, the very stuff we breathe to sustain life.

The difference between ozone and ordinary oxygen is in the number of atoms that make up the molecule. Ordinary oxygen contains just two oxygen atoms (written in chemical formulas as O_2), while the ozone molecule contains three of those same atoms linked together, written as O_3.

As it happens, a molecule which contains three oxygen atoms is a great deal less stable than the normal oxygen molecule with two. For that reason, ozone is far more chemically reactive than ordinary oxygen. It is for that reason that ozone, *at the Earth's surface*, is considered a serious pollutant: as we saw in the last chapter, that surface ozone comes largely from the action of sunlight on automobile exhausts, and its violent chemical reactivity seriously damages stone, rubber, vegetation and human lungs.

Ozone does not stop being very reactive when it is high up in the stratosphere, where the ozone layer is located, ten or more miles over our heads. The difference is that the air there is so thin that the ozone does not have much to react with.

Even so, individual molecules of ozone do not last very long even in the stratosphere before they do find something to react with and are destroyed.

What keeps the upper-air ozone layer in existence is that its supply of ozone is continually replenished by the Sun. The energy from sunlight breaks up ordinary oxygen molecules. Then some of them recombine in the three-atom grouping that is ozone.

There isn't really very much of that stratospheric ozone. Altogether there is no more than some five billion tons of it, which is nothing compared to the huge mass of the rest of our atmosphere. If that ozone were at sea-level pressures that would amount to a thin skin of gas no more than an eighth of an inch thick.

But that ozone isn't at sea level. Where the ozone layer lives, high up in the stratosphere, the pressures are so much lower that it, along with the other tenuous stratospheric gases, fills a layer twenty miles deep.

When we hear the words "ozone layer" we may form a mental picture of something like a shiny spherical shell of solid ozone, surrounding the Earth and reflecting all the bad ultraviolet-C back into space.

That isn't the way it works. What happens is that that same ultraviolet radiation from the Sun which formed the ozone in the first place now breaks the ozone molecule into two parts again. One part is a molecule of ordinary oxygen (O_2); the other is a free atom of oxygen (O). The free "monatomic" oxygen then joins spontaneously with another molecule of ordinary oxygen—perhaps even the one it was just split free from, though that will happen very seldom—to create another ozone molecule. In the process the ultraviolet-ray energy that went into splitting it in the first place is now released again, but this time in the form of heat.

The effect of all this is that the ultraviolet-C, which could harm us, has been transformed into ordinary heat, which can't.

All these transformations and exchanges are complicated. The purely chemical processes involved won't work properly except in the presence of certain catalysts—other chemicals which speed up a reaction or even make it possible in the first place. The "photochemical" ones—the ones driven by sunlight—are very strongly affected by the season of the year (the biggest Antarctic ozone losses occur in the months of September and October, which are the months of the Southern Hemisphere spring). Ice crystals must be present for the reactions to work; and all the processes are influenced, as well, by physical effects associated with changes in the circulation of the air, particularly those spiraling upper-air winds, located over the Arctic and Antarctic areas, which are called the polar vortex.

The effect of all that, however, is that by the time all those reactions have taken place, the deadly ultraviolet-C has been neutralized. The harmless heat that is left over simply radiates away into space.

Those are the rules of the game. As you can see, this game is actually a kind of race. The survival of the ozone layer is a competition between creation (by fusing monatomic and ordinary oxygen) and destruction (as the ozone is split up by ultraviolet-C). This race has been more or less a dead heat for billions of years. That had to be the case if we were ever to survive; if the ozone layer had failed to achieve a balance, life could never have developed on our planet.

But now we have interfered in the process. We have begun to manufacture synthetic gases which add to the destructive process, and now ozone is losing the race.

Of course, the ozone layer cannot be destroyed permanently. It will always keep coming back—as long as there is light from the Sun, and free oxygen in the air of the Earth— because the Sun keeps creating more of it.

But it takes time for the Sun to make new ozone, and if we allow the ozone layer to be seriously damaged, by the time it is regenerated we may all be dead.

That's where our interference in the ages-old race between ozone formation and destruction comes in, with the CFCs and other chemicals.

What we have done is to manufacture a group of synthetic gases, of which the most important are the man-made chemicals known as chlorofluorocarbons—CFCs for short.

The CFCs are not the only chemicals which attack the ozone layer. There are at least a dozen others, most synthetic, a few natural, which do the same thing. (We'll look at some of the others a bit later on.) But the CFCs are far more efficient at the process than the others, and it is they which are now threatening to destroy the ozone layer, and our very lives.

Scientists have known for a long time that CFCs presented a potential problem in this area. They had begun to suspect it in 1971, when, on a trip to the Antarctic, James Lovelock (yes, the "Gaia" James Lovelock) first discovered a surprising fact. His studies showed that almost all the CFCs ever manufactured since they were first synthesized in 1930 had not disappeared. They were still lingering in the atmosphere. Since they were so hard to destroy, they simply accumulated there.

A few scientists began to realize that this fact represented a real danger two years later, in 1973. That was when a scientist named F. Sherwood Rowland, working at the University of California in Irvine, carried out some experiments with them. What Rowland's studies showed was that the CFCs had the ability to act as extraordinarily effective catalysts in some of the gas reactions that destroyed ozone.

A catalyst, as we mentioned above, is a substance which takes a part in a chemical reaction—speeding it up, or even making it possible in the first place—without itself being used up. In that way, a single molecule of a CFC in the stratosphere attaches itself to an ice crystal. Then, when a random collision brings it in contact with a molecule of ozone, it destroys that ozone. And then, the CFC itself still intact, it can go on to do the same to another ozone molecule, and another, and so on almost indefinitely. It takes about fifteen years for CFC to migrate from its source in a chemical factory to the stratosphere. Then, once there, it is estimated that each molecule of a CFC will last for about a century, during which time it will destroy about 100,000 molecules of ozone.

Thus even a tiny quantity of a CFC can destroy many thousands of times its own weight in ozone before it itself is finally destroyed.

Rowland's research was all new science when it was announced. No one had ever conducted those particular studies before. No one had had much reason to do so, since until quite recently CFCs hadn't existed anywhere in the world. They are completely man-made.

They are also quite commercially valuable. Their greatest asset comes from the very fact that they are such stable chemicals, in fact almost chemically inert. They are not attacked by most other chemicals, nor do they attack them themselves. Therefore they are not poisonous. In fact, they have no known biological effect at all; in even very large dosages, they do no harm to living things. (This was demonstrated quite dramatically when Freon, the first of them to be commercially important, was put on the market. To prove that the chemical would do no harm, the scientist in charge publicly drank an entire tumbler of Freon.)

These CFCs have many uses. They are used to make foam plastics; the CFCs are injected into the plastics while still in the liquid state, whereupon they expand and create the foamy, bubbly, lightweight and heat-containing substances we use as plastic coffee cups, hamburger containers and packing materials. They are very widely used as the working fluid in refrigerators and air conditioners. They are the propellant in many spray cans for deodorants and hair cosmetics (still used for that purpose in most of the world, though that use is sensibly now prohibited in the United States). And they are used in the electronics industry for cleaning chips, circuit boards and other components, where the fact that they evaporate cleanly and leave no residue or chemical effects makes them ideal for the purpose.

Ultimately, all the CFCs do evaporate, whatever the purpose they were used for. Then they wind up as free gases in the air we breathe.

Breathing them in does no apparent harm, but when they diffuse up into the higher reaches of the atmosphere, particularly in the polar regions (where the ice crystals that facilitate the reaction are found), there is a different story. That is where they become ozone killers.

Remember how violently ozone reacts with other chemicals. It is just because ozone is so reactive that it, almost alone among reasonably common chemical substances, can react even with the generally inert CFCs. And in the process the ozone is destroyed.

The story of how the ozone problem was discovered is important—not to say *urgent*—in its implications for our future on Earth, but it's also interesting in another way. It's an example of the odd and sometimes ironic ways in which scientific discoveries are made.

Around the time when Rowland announced his discovery, in the early 1970s, there was a brief flurry of scientific worry about the ozone layer. Curiously, the worry was about the wrong things.

The report that CFCs could destroy ozone was noted, but that was not the worry. Fears about the safety of the ozone layer itself came from another source: airplanes.

At that time new generations of advanced supersonic jet aircraft were being designed or even actually manufactured—the Anglo-French Concorde was in production and the finally aborted American SST was still considered a possibility. Some scientific studies revealed that enough such aircraft flying at such a height might produce exhaust chemicals which would damage the ozone layer.

As it turned out, the Concorde was an economic bust and

the American SST never got off the ground. Whether high-flying supersonic jets would actually deplete the ozone layer hasn't been put to the test, because there haven't been enough of such planes to have any effect. (It might become a problem again if the "Orient Express" spaceplane beloved by Ronald Reagan ever got built, but that doesn't seem probable for the near future at least.)

But CFCs were not involved in this burst of concern. Jet aircraft do not produce chlorofluorocarbons. Anyway, although Rowland's work with CFCs was noted, there was no real proof that the laboratory reactions he described were actually taking place in the real atmosphere of the Earth.

Even so, just because Rowland had proven that such an ozone-destroying CFC reaction was possible some precautionary steps were indeed taken. Since 1979, the use of CFCs in spray cans (except for a few medical applications) has been prohibited in the United States, for instance. (Though their manufacture was not, and their use as spray-can propellants in most of the rest of the world has continued to grow unabated.)

But then public concern died down. Nothing was done to halt the growing, and now much wider, use of the CFCs in all their myriad other applications. The Carter administration had begun drawing up guidelines for other measures to limit their use—but Carter was beaten in his campaign for re-election, and the business-oriented administration of the incoming President Ronald Reagan was not interested in such do-gooder subjects. All the plans for further limitations were scrapped.

Then, in 1984, the scientific world got a startling shock. A separate group of scientists making measurements of the air over the continent of Antarctica discovered, more or less by accident, that there were holes in the Southern polar ozone layer.

The first discovery didn't really take place in 1984. It had actually happened in 1982; there just had been no public announcement then. The reason for the delay was caution. After all, there had been no warning of any real trouble with the ozone layer. The scientists had in fact been looking for something quite different. But there their findings were. The discovery was so unexpected that the scientists wanted to make sure before publishing, so they secured more sensitive instruments and repeated their observations, with great care, before finally announcing what they had seen.

So it was only in 1984, two years after the first observations, that the scientific world was told that, over a large section of Antarctica, two-thirds of the ozone layer had simply disappeared.

That caused more than a mere flurry. Other scientists took notice very quickly and started their own studies. An expedition to the Antarctic confirmed that first discovery, two years later still, in 1986; and in 1987 American aircraft, operated by the National Aeronautics and Space Administration out of Punta Arenas, Chile, systematically mapped the area of ozone destruction.

It had grown since 1984. By 1987, the disk of depleted ozone extended as much as ten degrees of latitude from the South Pole; within that circle, from 70% to 97% of the ozone had disappeared. Even for some distance outside the area nearly a quarter of the ozone was gone.

There was no longer any room for doubt at all: our protective shield against ultraviolet-C was being shredded.

There is a curious sidelight to this discovery. After the first findings were announced, some other scientists who were in charge of interpreting the observations of the American weather satellite, Nimbus 7, took another look at the stored recordings of what their satellite had reported as it passed over the South Pole.

Sure enough, the instruments on the satellite had faithfully detected the growing changes in the Antarctic ozone layer over a period of several years. But the scientists had not been aware of these findings.

The reason for their ignorance was that, as one colleague uncharitably put it, they had "shot themselves in the foot." They had simply done what good scientists always do, by taking a normal precaution against the possibility of recording false data.

Good scientists are always careful about such precautions, because occasional false readings are always a problem in complicated scientific observations. The best of instruments make wild excursions because of some external event, or simply because of the perversity of machines. So scientists regularly set their instruments to disregard transient errors.

That was what had betrayed the Nimbus scientists. Embarrassingly, they had programmed their instruments to reject any "impossibly" low measurements of the ozone layer, never imagining that a drastic loss of ozone might be quite possible after all.

So, by the mid-1980s, the concern was real and growing. All these developments gave a considerable boost to the stock of Sherwood Rowland in California, as his predictions, now a decade and a half old, turned out to be right on target.

Or almost right. There was one thing wrong with Rowland's investigations. They hadn't gone far enough; there were additional threats to the ozone layer which he had never considered.

As we know now, the CFCs are not the only man-made ozone destroyers. The world's chemical factories also produce great volumes of such other chemicals as carbon tetrachloride and methyl chloroform, and of the bromine-containing compounds called "halons," and all of these are also destructive to the ozone layer. (What's more, even some of the new chemicals that are being offered to replace these ozone destroyers—as we will see—are also damaging to ozone in the long run.)

So much for explanations. Now let's get into the meat of the matter: What's the worst that can happen to us because of this ozone problem?

The worst will be pretty bad. Worst of all, some of it seems to have begun already.

If this ozone destruction is allowed to go on, that solar ultraviolet-C radiation will do significant damage to life on Earth. It is not just a matter of more sunburns, and more painful ones (although that will happen), or of temporary "solar retinitis" or sun blindness (though that will destroy vision for many—particularly the young adults, who are not only the group most likely to over-expose themselves to the Sun, but whose retinas are not naturally as well protected as those of older people). A rule of thumb is that, in a community of a million people, for every 1% loss of ozone there will be 30,000 additional cases of skin cancer and 6,000 of cataracts.

If the ozone loss is great enough, the damage from the excess hard ultraviolet radiation will not even be limited to additional skin cancers and cataracts. The health of the whole body is at risk.

The human body's immune systems are weakened in a person subjected to excessive hard ultraviolet. These immune systems are what protect us from minor infections. When they are damaged infectious diseases of every kind can increase, from the common cold to leprosy, as the natural defenses against disease lose some of their effectiveness. (The recent plague of AIDS, although it comes from quite different causes, is an example of what this loss of immune response can do to a victim.)

We human beings can at least partially protect ourselves against some of the effects of increased ultraviolet radiation. We can make it a habit to wear hats and sunglasses when we go out. We can daub ourselves with sun-blocking creams. Most effective of all, we can simply stay indoors as much as we can. We might even be able to keep some of our domestic animals in the shade, at least to some extent, though that gets very difficult even to imagine—who is going to spread sun-blocker cream on tens of millions of cows? But the rest of Earthly life doesn't have that option. Perhaps the most finally dangerous effect of excessive ultraviolet-C radiation will be what it does to the world's vegetation.

Plants, which comprise 99% of the mass of living things on Earth, by definition cannot live without sunlight. That's what a plant is, an organism that survives by tapping the energy of the Sun to grow.

If UV-C increases too much, some plants will simply be scorched to death. Other kinds of vegetation, including many of our food crops, will be damaged in a variety of other ways.

Some of these other kinds of harm are not immediately obvious, such as the crop damage that will come from lowered soil fertility in, for example, rice paddies. (In rice cultivation a group of microorganisms called cyanobacteria add fertilizer for the rice crops by fixing nitrogen. The cyanobacteria are very valuable to the rice farmers in this respect—cyanobacteria fix some 35 million tons of nitrogen every year, which is more than all the nitrogen-bearing fertilizers manufactured in the world supply—but they are very easily damaged by ultraviolet rays; and without them crops will dwindle.) The effect of all this damage produced by weakening of the ozone layer will be felt in lowered yields from the world's farms—and ultimately in crop failures and famines.

Excessive hard ultraviolet even does serious damage to many things which aren't living at all, as well: plastics, textiles, paints and rubber all have their useful lives shortened by such exposure.

If the ozone layer disappeared completely, all over the world, we could not survive. Of course, the whole layer isn't likely to be destroyed entirely, no matter what we do, since, as we have seen, new supplies of ozone are continually re-created by the sun. But a worldwide, year-around loss of 15% is possible—indeed, it is predicted by some authorities to occur within the next few decades—and that is quite enough to cause widespread suffering and deaths.

Now perhaps we want to do some reality-testing again, beginning with the question: "Is this all real?"

It is certainly real, although obviously the worst of the anticipated effects have not shown up yet.

That that is so is understandable enough, because what has happened to the ozone layer so far has generally been happening in places where there are few living things to suffer.

Antarctica is just that sort of a remote and barren place. It is also where the ozone loss was first discovered, because the Antarctic spring gives just the right conditions for ozone destruction. Therefore it is the first place for us to look for evidence.

No dramatic die-offs have yet been observed in the Antarctic. That is reassuring, up to a point, but it would be a lot more so if it weren't for three facts:

First, compared to the rest of the world, there isn't much life there to be damaged, in those frigid and nearly sterile South Polar regions.

Second, we might not know very much about die-offs even if they had occurred, because the necessary studies to see just what *has* happened to the creatures of the Antarctic area, like so many other investigations, are actually just beginning to be made.

Third, most Antarctic life is somewhat sheltered from any effects related to sunlight by a cloak of water. Most of the organisms that live there spend much or all of their time in the sea and sea water screens out a good deal of the Sun's radiation. Much of the Antarctic's life consists of fish and smaller organisms that live in the ocean full-time. Among the air-breathers, the only important exceptions to this rule are the flying birds.

The rest of the air-breathing life of the Antarctic comprises a few marine mammals—whales; but of course most of the whales have long since been hunted to near extinction anyway—but mostly such non-flying birds as the Adelie,

chinstrap and gentoo penguins—more than a hundred million penguins altogether—and the animals that live primarily by eating the penguins, such as the leopard seal. All of
these creatures are forced to spend much of their lives in the
sheltering water, for it is in the water that their food is to be
found.

We need next to look at what that food may be. For the
penguins, their food source is almost exclusively krill. Krill
are tiny shrimplike creatures, the largest of them less than
three inches long. They swarm in the Antarctic Ocean in
vast numbers—at the height of the Antarctic summer, as
many as 650,000,000 tons of them—and they are a great
food resource for the larger animals of the region. Until intensive whaling in the middle of this century nearly exterminated the Antarctic baleen whale population, the whales
ate great quantities of the krill. Now the krill feed other
marine predators and the penguins. Evidently they feed
them very well, since there appear to be more penguins
alive in the Antarctic today than ever before in the history
of the world, thriving on the extra food the whales no longer preempt. In addition, krill are now harvested regularly
for human consumption (mostly as fish meal) and for animal feed. The krill fisheries, operated mostly by Japan and
the USSR, are fairly new and still relatively small—at the
height of the season, the catch may amount to no more than
100 tons a day. The krill have the promise of becoming a
more important food resource for human beings at some
future date, but at present there are technical difficulties yet
unsolved that get in the way of their being readily accepted
for human food.

No evidence has yet been found of ultraviolet damage to

the penguins and seals, but there are already some indications that the krill catch may be dwindling.

Is that because the Antarctic's ozone layer is letting more of the deadly ultraviolet through? It's hard to be certain of that, because there are too many unknown factors; here, too, the studies that would give a reasonably reliable answer are just now being made. But there are some things we do know as facts. We know that the food the krill live on is the microscopic plant life of the Antarctic Ocean. As is true of all plants, those tiny marine plants require sunlight. They can't carry on the photosynthesis that keeps them alive without it and therefore *must* expose themselves to whatever radiation the sun may provide. And we know that, in recent years, more and more of that Antarctic sunlight has contained a deadly component of hard ultraviolet.

Scientists now believe that the excess ultraviolet-C is driving these tiny planktonic plants deeper into the sea. That gives them some shelter from its effects. It also, though, results in their growth being stunted by lack of the beneficial parts of the solar light. (Small marine plants of similar kinds, by the way, also do a significant proportion of the vegetable kingdom's work of removing carbon dioxide from the air worldwide—so when they are damaged the greenhouse effect goes even faster.)

So much for the Antarctic. There is certainly cause for some concern, but after all the Antarctic is quite remote from the lives of most of us. Unfortunately, the ozone hole over the Antarctic is not likely to remain just where it can do so little harm to living things.

As ozone is destroyed, the depleted parts of the ozone layer must necessarily grow in size; as upper-air winds work on them, some of those depleted patches of ozone will break away and be driven to migrate farther northward, to populated areas.

They are known to have already begun to do so. By the time of the 1989 Antarctic spring as much as 15% to 30% of ozone was lost as far outside the previous boundaries of the hole as 50 degrees south latitude.

That's coming into human-habitation country now; that 50-degree parallel cuts across the extreme southern tip of South America and comes close to Australia and New Zealand. Sure enough, parts of Australia have already detected higher than usual ultraviolet radiation at times when migrating clots of ozone-poor air drifted north.

Nor is it only the southern hemisphere that is affected. Smaller, but similar, ozone depletion has been observed in the North Polar region as well.

And here in the northern hemisphere some of the effects have reached spots definitely inhabited by quite large numbers of people. In Sweden scientists have registered lowered ozone concentrations for some time (though, curiously, Norwegian scientists have not), and Soviet scientists have measured thinning of the ozone layer over several of their cities. In the Swiss Alps, ultraviolet radiation has been measured to increase about 1% since 1981. Presumably this is due to ozone losses, which are estimated to have amounted to a 5% depletion over much of the northern hemisphere of our planet during the winters of the 1980s—in some patches, as much as 17% of the ozone disappeared.

Cases of actual physical damage to human health have already begun to turn up. In North America, as far south as Michigan and New York, there was a temporary increase in cases of sun blindness in March, 1986, as a patch of depleted atmospheric ozone moved across the northeastern part of the country.

Once again we must ask ourselves if this is all *real*, for it is a fact that, in spite of all we have described, a few scientists are not convinced.

These skeptical scientists do not deny that CFCs can destroy ozone, or that ozone is being destroyed. They merely argue that the exact nature of the connection between CFCs and the undeniable tattering of the ozone layer has not yet been proved, and so they suggest that eliminating the manufacture of CFCs may not save the ozone.

These scientists point out that the complex chemical dynamics of the polar stratosphere are still poorly understood. They are quite correct in saying that, too. Certainly more factors are involved than the reactions between the ozone molecules and the CFCs alone. Among those other factors are such imperfectly predictable matters as the amount of ozone-creating solar radiation, which varies slightly from time to time; the complicated Antarctic wind and weather patterns, which determine how much mixing goes on in the stratosphere; the abundance of ice crystals on which the CFCs collect; and no doubt many other things.

But even if this minority view should happen to turn out to be right and other factors play an important part, it wouldn't make any difference to what we must do. We would still have to do our best to outlaw production of the CFCs. They certainly do play some part in the destruction of ozone. And besides that, there is one important other thing that cannot be denied:

Every other possible element implicated in the destruction

of the ozone layer—solar fluctuations, air circulation, etc.—is securely out of human hands.

If by some improbable chance one of those other factors should turn out to be significantly important in the ozone loss, there would be nothing we could do about it. We would simply be out of luck. The ozone would be destroyed, the deadly ultraviolet radiation would survive to reach the Earth's surface and we would have no choice but to suffer the disastrous consequences.

So the CFCs and the other destructive chemicals are not only overwhelmingly likely to be the most important cause of the destruction of the ozone layer, they are the only ones that we have any means of controlling.

The answer to the question of whether this is all real is "Yes." The destruction of the ozone layer by the chemicals we manufacture and pour into the air is real, and so are its consequences. The problem is becoming more serious every day. It will not get better very quickly, either. If we stopped producing CFCs today—which is not going to happen—there would still be significant ozone loss continuing well into the twenty-first century for much of the world . . . and in some places, like the Antarctic, a detectable ozone loss would still exist for as much as two hundred years later than that.

After our discussion of the global greenhouse-effect warmup we tried to answer two other questions, the first of which was, "Can natural processes save us here?"

In the case of the ozone layer, the answer is that they can't. "Gaia" has no power to save the ozone. There are no natural processes that can do anything useful to remove chlorofluorocarbons from the air, since the chlorofluorocarbons are so extraordinarily chemically stable that they possess only one natural "sink" in the world's ecological checks and balances.

Unfortunately that sink is the very process we want to prevent. It is the slow destruction of the CFCs that takes place in the stratosphere, in the course of their much more rapid destruction of the ozone.

So the answer to that question is flatly, "No." Which leaves only the final question—"Is there anything we can do about it?"

This time the answer is, "Yes. Absolutely. We can stop manufacturing the chemicals that destroy ozone. That won't prevent damage to the ozone layer, but it will keep it from getting far worse; and the sooner we do it all over the world the better."

Our natural first step will be to calculate an ozone SAP. (You remember that, since most of the damage we are doing to our world takes the form of a race between destruction and healing, we can therefore try to calculate that level of destruction which the healing processes will be able to compensate for—what we have called the Steady-state Allowable Perturbation, or SAP.)

It's possible to do this for the destruction of the ozone layer. Since the Sun keeps making more ozone at a fairly steady pace, all we need to reach a balance is to cut down the speed with which the ozone is destroyed—by, for instance, limiting the amount of those ozone-destroying gases that enter the atmosphere.

That gives us some numbers to start with. The present concentration of chlorine (the part of the CFCs that does the damage) in the atmosphere is about three parts per billion. That's only a temporary figure. It will not stay that low. In fact, it is expected to increase steadily, reaching four parts per billion by 1997, even if the current international agreements limiting production are enforced (which is by no means a prospect that we can bank on).

Both of those figures are too high for our best interests. The maximum sustained level which would produce only minor damage to most of the world has been calculated to be slightly less than *two* parts of chlorine per billion parts of air.

So our SAP must therefore limit production of the chlorine-containing synthetic gases to a level which will cut the amount of chlorine in the air at least in half.

Do we have a number for that production level? The author of *A Hole in the Sky*, John Gribbin, offers one rough estimate: If we cut worldwide production of CFCs down to about a sixth of the 1988 levels, that would at least keep our prospects from getting worse than they already are.

Note that word "prospects," though. We're not talking about preventing our *situation* from getting worse. The most we can hope for is to limit the increasing *future* damage to a level we can somehow stand.

We must always keep in mind that, as we have mentioned before, the damage to the ozone layer will inevitably get somewhat worse before it can begin to get better, no matter what steps we take now. It's already too late to prevent that. The assassins of the ozone have already been set free; they simply haven't found their victims yet.

Most of the CFCs that have been manufactured already are in existence, floating around in the lower parts of the atmosphere all around the world. There is no imaginable "technological fix" by which these existing CFCs could be found and destroyed—they are too sparsely spread, and too chemically sturdy. Nor is there any way to prevent them from slowly diffusing upward into the stratosphere, where they will meet the ozone and destroy it. Little by little, the gases we have already manufactured will keep on doing this, no matter what we do, for many years into the future.

In any case, that doesn't accurately represent the realities of the situation. The manufacture of CFCs hasn't stopped. It hasn't even slowed down.

There have, it is true, been some attempts at legislating a curb. The United States Environmental Protection Agency issued an order in 1976, under the Carter administration, to prohibit their employment as a propellant in some kinds of spray cans. That was a good beginning. A beginning was all it was meant to be; the intention was that that would be only Phase 1 in a more thoroughgoing series of regulations on CFCs to be put into effect later, but Phase 2 never followed. Ronald Reagan was elected, and his choice for EPA administrator, Anne Gorsuch, dismissed the whole problem as just another environmental fantasy.

In consequence, worldwide, there are more CFCs being produced now than there were in 1976. It wasn't until 1979, for instance, that one of the principal demands for CFCs arose. It was then that the electronics industry took up their use of CFCs as solvents in a big way; and, because of those new uses and because most of the world still is expanding its consumption through old uses, the production of CFCs is still actually increasing.

One would think that, knowing all this, some steps would have been taken long ago to avert this thoroughly predictable catastrophe.

Actually, of course, as we know from having read the newspapers, some steps have indeed been taken. The principal ones are the international agreements we've already alluded to.

These agreements came about in 1987, when a world conference on the threat to the ozone layer was convened in Montreal, Canada. On September 17th of that year the attending delegates from twenty-seven nations signed a treaty to protect the ozone layer. Under it, they promised to reduce their aggregate production of CFCs and one or two other ozone-destroying gases by 50% by the year 2000.

Of course, this agreement is much too little, and certainly quite a lot too late for its best effects. Such a step ten years earlier might have made a difference; now the global atmosphere is already flooded with ozone killers. Nor does that 50% cut come anywhere near even John Gribbin's already fairly permissive suggested production cut of five-sixths. Nor does it obligate in any way the hundred-odd other countries of the Earth, who were not represented and who have made no commitment to limit their production of the ozone-destroyers at all.

Finally, even that inadequate treaty covers the manufacture of only a fraction of the chemicals which are known to destroy ozone.

We've talked about the CFCs as the prime culprits in the killing of the ozone layer. They are that, but they are certainly not alone. There are other anthropogenic compounds that have similar effects on the ozone layer—for example, carbon tetrachloride and methyl chloroform—and they are not dealt with in the Montreal accord. Worse than that, some of the "safe" substitutes for the CFCs—for example, the classes of chemicals known as the hydrochlorofluorocarbons (or

HCFCs—compounds related to the CFCs, but with hydrogen added) and the hydrofluorocarbons (HFCs—again related to the CFCs, but with the chlorine removed and replaced by hydrogen) are not only not prevented by the Montreal agreements, they are actively recommended as replacements for the prohibited ones.

There was a certain reason—or at least a hope—behind this decision in Montreal. The rationale is that the HFCs have little or no effect on ozone (though they can't replace the CFCs in many applications, at least not without considerable problems of design and manufacture), while the HCFCs, although they do attack ozone, are only about a tenth as effective at it as the CFCs.

A 90% improvement is certainly *better*. Equally certainly, it is not good *enough*. Unrestricted production of these "benign" chemicals will inevitably mean producing so many of them that their comparative weakness in effectivity will be overwhelmed by their quantity in the air . . . and so our grandchildren will be no better off. (In fact, two of the largest CFC manufacturers, Du Pont in America and ICI in Great Britain, have already begun constructing new plants for the HCFCs. Half a dozen such plants, in Canada, the USA and the British Isles, will be producing nearly half a million tons of the HCFCs by the time this book appears.)

Did the delegates in Montreal know this?

Of course they did, or at least most of them did. They knew more than that, too. They knew that there was no reason for synthesizing any of these chemicals, because *nearly everything done by the CFCs and their suggested synthetic replacements can be done very nearly as effectively by naturally occurring substances that do no ozone damage at all*—by water, for instance, or by such inert gases (in such applications as refrigeration and air-conditioning) as helium or carbon dioxide.

What they also knew, though, was that their governments back home were at least as worried about economic harm as about the harm to the ozone layer. The trouble with replacing CFCs with, say, water—from the point of view of the chemical industry—is that they can't sell water; and their voices in the ears of government were far louder than those of the environmentalists.

All of this certainly sounds discouraging. It is not utterly hopeless, however. We wouldn't have bothered writing this book if it were.

At the present moment, the exact dimensions of what we have to look forward to in the future are unclear, because that particular future hasn't been "invented" yet.

We're all inventing it right now, all of us in this world. As the world learns what sort of perils it faces, one would like to think that it will try to invent a more benevolent one than that which faces us right now. That could make things considerably better. If we—the worldwide "we" that includes the whole human race—take the proper measures, our grandchildren may enjoy sunlight that is very little different from our own.

If we don't, their future is far more bleak. They will occupy a world where many plants cannot survive in the open sunlight; where human health is damaged, not only by the immediate skin cancers and cataracts caused by ultraviolet-C but by such indirect ones as an increase in almost all diseases as ultraviolet-damaged immune systems lose their ability to protect the body . . . where even the soils we grow our food on will be seriously damaged . . . even a world where such sturdy items as plastics, wood, rubber and textiles may be embrittled and aged by that exposure.

They will, in short, inherit a savagely damaged planet on which to try to live.

How damaged will our planet be? At the extreme, Carl Sagan has pointed out that we can get a pretty good idea of what an Earth without any ozone layer at all would look like simply by studying spacecraft photographs of the bleak and lifeless planet Mars. As Sagan puts it, "The antiseptic martian surface is what you get when your ozone goes away."

It obviously behooves us to invent a better world than that.

It would be nice to stop this catalogue of man-made disasters here. We've already established that we human beings are endangering the world's weather, poisoning the world's air and releasing floods of damaging radiation on the world's life by weakening the ozone layer; do we really need to hear any more?

Sadly, we do. Everything we are doing to the air we breathe we are also doing to the water we drink and the soil we grow our food on—and more—and, not content with polluting the Earth we inhabit, the astonishing truth is that we are even going on to pollute the space around us.

So let's grit our teeth and complete the bill of indictment against ourselves.

We promise that when that is through we will come to more hopeful topics: There *are* solutions to all these problems. But before we get to the cure we must finish with the diagnosis. For that we must look at our other assaults on the world we live in, starting with our reckless poisoning and waste of the world's water.

10

The Water We Drink

When we talk about what is going wrong with our water supplies, it has the same old-fashioned sound as talking about the pollution of our air. They both sound like the *last* generation's ecological worry.

Unfortunately they are both still with us. For both, a good many expensive and laborious cleanup projects have been undertaken. In some localities some of the attempts have shown some partial success. Nationwide, and above all worldwide, they have failed. As we have already seen, even the cleanest American air is tainted with acids and smog-producing chemicals, while the water we depend on for life is generally dirtier—and all too often it is scarcer—than ever.

That's dismaying—no, it's worse than that; it is *scandalous*—when you consider the realities of the situation. It is a fact that, apart from its dry Southwestern states, the United States is unusually blessed by nature with plentiful supplies of fresh water. It contains the Great Lakes, one of the largest bodies of fresh water in the world—only Lake Baikal in Siberia holds more. Great volumes of fossil water underlie much of the country, available for people to use for the price of sinking a well; and its rivers are vaster and more numerous than anything in Europe. With the least prudence, there could be no such thing as a water problem in the United States.

But that kind of prudence has been conspicuously lacking for more than a hundred years.

One reason for our water problems is that we build great cities to hold our people in places where there is not enough natural water to supply them. Southern California is the textbook example. There is a certain amount of natural water that is locally available for the city of Los Angeles—that's why the city first grew there. That local supply is not small. It is about enough to meet the needs of a million people or so—but it is hopelessly inadequate now that the area has exploded to contain *fourteen* million people. That is why Los Angeles has already reached out its aqueducts to suck into itself almost everything that flows in its own state and nearby ones, and now has even begun to cast a thirsty eye at the vast Canadian rivers two thousand miles away. (Canadians are not at all enthusiastic about the idea of pumping their water south.)

Everybody jumps on Los Angeles, but in this case it and the other California cities are by no means the only offenders in that state. California's farmers are actually worse. In all, California's agriculture uses about 85% of the state's water. That might seem fair enough—farmers must grow crops so people can eat, after all—but a huge amount of that precious water is wasted on such lunacies as irrigating pastureland for dairy cattle; it is an astonishing fact that watering grass for pasture swallows more water every year than the cities of Los Angeles and San Francisco combined.

Still, the California cities are bad enough, with their lawns, fountains and jacuzzis. Some of the cities in nearby states are even worse. The bone-dry states of Arizona and Nevada have plunked down large cities where there used to be nothing but cactus. Las Vegas—which has no discernible purpose for existence except to provide a gambling den—is studded with acre-sized pools, fountains, decorative ponds, sprinkled lawns and more than twenty irrigated golf courses; every drop of the water that makes the city green is imported from the already hard-pressed Colorado River. What's left of that river after everyone else has taken a turn at it is not enough for a city of 800,000 people like Las Vegas, with the number growing every day. So Las Vegas is now trying to get permission for a five-billion-dollar system of aqueducts with which they intend to pump in the fossil water from the aquifers under the northern ranch counties of Nevada. (The ranchers, like the Canadians over their rivers, don't want that plan to go through. There's a little bit of comedy in their situation. For years the ranchers have been fighting the efforts of such organizations as the Sierra Club to turn some of those areas into wilderness preserves; now they are begging the conservationists to come in and help them against the Vegas casinos.)

Well water, too, is becoming both dirtier and scarcer all over the country. Well water comes from the underground water supplies called aquifers. They are not infinitely large.

We speak of aquifers as though they were underground storage tanks, like swimming pools covered over with soil. Actually, an aquifer is a kind of sandwich made up of three different layers of rock or soil. The top layer may be clay, pumice or some poreless rock like granite; this forms a cap because water can't circulate where there are no pores or where, as in pumice, the pores don't connect with each other. The bottom layer of the sandwich is also impermeable; this keeps the aquifer's water from draining away.

The middle layer is where the water is. This layer—the meat in the sandwich—is gravel or some porous rock like sandstone. This layer holds the water in the spaces between the solid material, and allows it to flow from one point to another.

Therefore, if the top layer of the sandwich is open to the elements anywhere over its area, rainwater can enter there. Then the spaces in the gravel or sandstone can fill with rainwater, and the water seeps all through the aquifer. Then if you dig a hole through the top layer of the sandwich into the aquifer "meat" in the middle, you have a well.

It doesn't matter where in the aquifer you dig. Your well can even be hundreds of miles from a replenishment point. The water will flow from all over the aquifer to replace what you take out.

You may have to dig pretty deep to strike water—as much as half a mile sometimes—which means you have to spend a lot of energy in pumping the water up to where you can use it. (There is some water even farther down, but at those depths getting it to the surface is impractical, if not hopeless.)

Contrariwise, you may find that the water comes fountaining out to the surface with no pumping at all. This is called an "artesian" well. It comes about when the replenishment point is higher than the surface of the land where you dig your well. Since water seeks its own level, the water will rise to run out the top of the well—or may simply appear as a natural spring. (In a few places, notably around Minnesota, some farmers found artesian wells with so much pressure that they ran the outpouring water through turbines to provide energy.)

So finding an aquifer under your feet is good news. There are two flies in the ointment. First, the water in the aquifer may be salt, or it may be tainted by industrial or other chemicals. Second, underground water is fossil water. It may have taken thousands of years for the aquifer to fill. If you now pump faster than it can recharge—especially if you have a lot of other people pumping from the same aquifer at the same time—it will sooner or later go dry.

On a small scale, Las Vegas's thirst for water of all kinds may produce such a problem, for it is threatening not only the aquifers that serve Nevada's ranchers, but the ones that supply springs as much as 250 miles away—even the wildlife in Death Valley National Monument may go thirsty as that area's aquifers are drained down by Las Vegas's demands.

More serious to the country is the condition of the Ogallala aquifer. That aquifer underlies many Western states, from South Dakota to Texas and New Mexico, and it is one of the greatest sources of underground water in the world. It is the principal source of irrigation water for thousands of farms; in fact, it supports 20% of America's irrigated farms. Water is being drawn from the Ogallala fifty times as fast as it can be replenished from rainfall, and so the immense aquifer is being rapidly drained.

In parts of the Southwest the water table of the Ogallala has fallen by dozens of feet in the lifetime of people now there; in many places there is no water left at all, and rusting pipes are all that remain of the irrigated circles that once dotted the desert. At present rates, by the year 2020 there will be no Ogallala water at all left under New Mexico, and greatly diminished amounts almost everywhere south of Nebraska.

In Florida (and many other places in the United States and the rest of the world), excessive pumping has lowered the underground water levels near the sea shores so much that salt water is seeping into the aquifers. While in many, many places where there is plenty of water nearby seepage of chemical and industrial pollution has ruined underground wells for householders: Love Canal, almost within the sound of the huge volume of water that rushes over Niagara Falls, is the classical example.

Surface water supplies are almost as endangered. Even North America's greatest river, the mighty Mississippi, is showing signs of severe strain.

From its source at Lake Itasca in Minnesota to the muddy delta it has created in the Gulf of Mexico, the capacity of the Mississippi River has been stretched to the limit in all of its uses, for drinking, for transport and for irrigation. Much of its water is not only used, but re-used and re-used over again. Cities which get their urban water supplies from the Mississippi—including St. Louis and all points south—have to reckon with the fact that as much as 75% of their municipal water has already passed through the water systems of upriver populations.

Most of the people in those cities prefer not to reckon with the other fact about their water supplies: that is, that much of that same water has also, of course, passed through all those upstream kidneys. That may be a considerable esthetic problem for sensitive people. Fortunately it isn't a very serious medical one. The purely biological contamination that comes from the sewers that drain into the Mississippi can generally be successfully dealt with by municipal water-purification systems, as we'll see a bit later.

There is, however, a real health problem in using Mississippi River water, and that is its load of contamination by industrial waste. That is beyond the capacities of most municipal water systems, which do little to cope with industrial pollution. By the time the Mississippi water gets to Louisiana it is already carrying a burden of such poisonous wastes as PCBs and heavy metals. In Louisiana is where the real assaults on the river's water begin. The entire 150-mile stretch of the state between Baton Rouge and New Orleans draws on the Mississippi River for municipal drinking water. At the same time, the 130 oil refineries and chemical plants that line those banks use the same river to dispose of their waste products, which include such toxic chemicals as vinyl chloride and benzene. The combination spells poison. Although many of the people in the area who can afford it drink only bottled water, that part of Louisiana has one of the highest rates of deaths from cancer of lung, stomach and gallbladder in the country; for those who drink river water (rather than well or bottled water) there, their risk of rectal cancer doubles.

The Mississippi is not only a source of drinking water, it is a main thoroughfare for commerce. In a dry summer like the drought year of 1988, that waterborne shipping literally dries up. The lack of rain means that the river's feeder streams have less water to carry into the Mississippi. When that happens the whole river begins to shrink. In a major drought like 1988's the level of the river drops so low that shoals turn into islands, channels become shoals and stretches of the river are no longer navigable. Then the thousands of barges in the Mississippi fleet, with all the essential cargoes they carry up and down the river, are stranded wherever they happen to be caught, sometimes for weeks on end.

The water level in that greatest of American reservoirs of surface water, the Great Lakes, has not dropped consistently in the same way. In the late 1980s it mysteriously went in the other direction: the water level of Lake Michigan actually rose so much that in storm surges it threatened lakeside buildings in Chicago. No one knows exactly why, and it has receded again since; there is evidently a long-term, and poorly understood, cycle of rising and falling of the lake level. In the event of the expected global warmup the level will probably recede somewhat more because of lessened rainfall and increased wind evaporation, but hardly to the point of disappearing.

Nevertheless, vast as they are, the Great Lakes are seriously polluted—in the case of Lake Michigan, to such an extent that pregnant women in that area are advised not to eat lake fish more than once a week. Such warnings are sadly common in the world today, but in this case the source of the pollution is surprising. A good deal of the current contamination still comes from its traditional source, the wastes of shorefront industry, but recent legislation has succeeded in cleaning a good deal of that up. The worst threat now is airborne. Half the PCBs in Lake Michigan, and 90% of those in Lake Superior, are blown in from industries and other sources hundreds or even thousands of miles away.

It would cost an estimated $100 billion to clean up the Great Lakes. (For comparison: all the cleanup efforts of the past twenty years combined have amounted to only a tenth that much.) And none of this cleanup planning allows for the quite likely worsening of the water contamination through future shipping accidents.

After all, the Great Lakes, like the Mississippi River, are a major transportation facility. They carry an immense volume of shipping from all over the world, by means of the

St. Lawrence Seaway's access to the ocean. Among those vessels are a good many oil tankers. If Lake Michigan, for example, should suffer the kind of near-shore oil spill that we have seen in Alaska, in the Gulf of Mexico and in many other places around the world, the drinking water supplies for some of America's largest cities would be dried up at once. There are no alternative sources that are anywhere near adequate; if you would like to stretch your imagination, consider just how one might go about beginning to try to solve the problems involved in attempting to truck sufficient water in to meet the needs of a city like Chicago.

Almost all problems are global problems these days, and the problem of scarce or polluted water is not by any means confined to the United States.

In France, the dry year of 1989 caused some French river levels to drop so far that atomic-power plants along their margins had to shut down for lack of water for their cooling systems. The main rivers of Europe, the Danube and the Rhine, have been turned into open sewers; the Dutch at the mouth of the Rhine, in Rotterdam, have to dredge out thousands of tons of polluted sediment from the bottom of the river each year (and then find a place to dispose of the sediment), while the "beautiful blue Danube" can no longer be called blue—or even beautiful; in places it is poisoned with the wastes of endless factories and in others it simply stinks, since such cities as Budapest along its banks flush all their sewage, untreated, into the river. Germany's Elbe River is even worse, with ten times as much mercury and other toxic metals as even the Rhine. It was into the Elbe, at the end of World War II, that American and Soviet soldiers waded to greet each other as they completed their pincer drive through Germany. They would be risking their health to do it now, but it still supplies *drinking water* to many communities along its banks. So does that other river in what used to be called East Germany, the Saale, although the Saale's pollution is such that, along half its length, even the most expensive and sophisticated purification procedures available can't salvage its water. Even the Alpine rivers are now contaminated, though in an unusual way. When snowfall fails, as it did in three successive years at the end of the 1980s, the Swiss don't want to disappoint their ski tourists; therefore they spread salt on the glaciers to make a passable skiing surface, and the salt, of course, winds up in the previously pure glacial streams.

Again, the farther east we go in Europe the worse the pollution gets. Most of Eastern Europe has similarly despoiled its water. In the former East Germany open uranium waste dumps pollute groundwater in the south, and industrial wastes damage water all over the area; only 3% of their lake water is drinkable. In Poland, it is estimated that 10% of the country's GNP is lost because of pollution, largely of water too filthy to be used by industry. In Czechoslovakia 70% of rivers are polluted, and 40% of all sewage goes into the water systems without any treatment at all. In the Soviet Union itself the Aral Sea is drying up because of diversion of water from its rivers to irrigation, while even Lake Baikal is losing some of its most prized food fish to pollution. Baikal is a mile deep, for which reason it exceeds even the Great Lakes as the world's Number One aggregation of fresh surface water. Until recently its pristinely pure water was so valuable that the Soviets located many chemical plants near its shores. For instance, in the production of high-quality rayon threads nearly chemically pure water is necessary. It was found, to the delight of the engineers, that Baikal's pure water could be pumped right out of the lake and used without needing any treatment at all . . . until the factory's own discharges so contaminated Lake Baikal that expensive water-purification facilities had to be built to save the production process anyway, just as they would have been if the plant had originally been sited in a less ecologically vital area.

In fact, all of the USSR's water is under threat to one degree or another. Because so much water has been diverted for irrigation, the Aral Sea has shrunk—vast desert areas border its present shores—and the salt content of the sea itself has risen from 10% to 25%. This has directly affected drinking water supplies for people nearby. Much of it is so polluted that even boiling doesn't make it fit to drink. Nearby people freeze their water into ice cubes, then crack them open to pour out the concentrated pollution of salt and pesticides in the center, melt what is left of the cubes and drink it.

The Indian subcontinent has all of everyone else's problems, and it also has a few that are uniquely its own: One of the sources of pollution in the Ganges River comes from the Hindu religion's practice of burning their dead. Since fuel is costly, the corpses are not always completely burned, and what is left is thrown into the Ganges. There they decay and despoil the river. India's present stopgap attempt at a solution is to breed 7,000 carnivorous turtles, which they intend to release into the river to eat the corpses. (Corpses are a problem for water elsewhere, too. In England some small traces of formaldehyde in water supplies are suspected to come from the embalming fluid used to prepare corpses for interment in cemeteries.) In China, the aquifers are being depleted so rapidly that the water table under the whole great South China Plain is dropping at the rate of five feet a year.

In the Middle East, Egypt is totally dependent on its sole source of fresh water, the Nile River, and it has already pressed the river to its limits. Now Egypt is threatened with increasing water scarcities as countries nearer to the Nile's source are developing their own uses for its water, and the flow that reaches Egypt's thirsty farms is thus decreased. As we noted earlier, concern over the Nile flow has led Egyptian politicians to threaten war against upstream neighbors who endangered it. In a separate project, Egypt planned to join the Sudan in digging a great canal through the Sudd marshes, far upstream, where much water simply evaporates as it trickles slowly through the wetlands; but revolution in Sudan put a stop to the effort while it was still incomplete.

In sub-Saharan Africa, Nigeria had a plan to irrigate much of the drought-stricken Sahel with water from Lake Chad in the 1960s. Even greater droughts ruined that project. Lake Chad has shrunk; the fifteen hundred miles of canals are empty, and rotting hulks of ships now lie high and dry, as much as thirty miles from the new shores of the lake.

And so on and on; freshwater systems all over our planet are being strained and, worse, a lot of the fresh water is no longer really fresh as industrial pollution, acid rain, agricultural runoff and sewage pollutes more and more of it. Worldwide, the situation is tragic. Because of shortages and pollution nearly two billion people—four out of every ten persons alive—drink and bathe in contaminated, disease-carrying water.

So, in spite of all the talk and effort, the water problem is not yet solved.

Nor do we have any choice about whether we solve it, either. We can't get along without fresh water. We have to have it for two main purposes—well, maybe for three. The two that are inarguable, because they are a matter of survival, are, first, to drink (and to use for bathing, washing things and so on) and, second, for agriculture, to irrigate our farms.

The third one is mostly esthetic, and so it only matters if you count swimming, fishing, boating and other forms of water-related recreation and pleasure as a real need rather than a dispensable amenity. But in some ways that is the easiest problem to solve, so let's talk about it first.

Water is *nice*. We may not urgently need, but most of us certainly do desire to have, water available to us for pleasure: rivers to fish in, lakes to boat on, bodies of water to swim in or simply to admire for their beauty. We are even willing to pay a substantial premium to live near open water, if only to be able to look at it out our windows.

Nature—helped by Gaia, if you like—has provided us with all those things in large quantities . . . but of course we have spoiled many of them with pollution.

The specific kinds of pollution that diminish our pleasure in our bodies of fresh water come from four main sources: the famous acid rain (and other airborne pollutants); industrial wastes; agricultural runoff (of excess fertilizers, pesticides and so on); and sewage.

These human assaults do not all affect the water in the same way. Paradoxically, while there are many lakes where we have killed off the fish by making the water sterile, there are others where we have killed them just as dead by making the water too fertile.

That kind of water pollution is the process called "eutrophication." It is caused by the runoff of fertilizer from agricultural lands, and from the sewage (including the phosphates from household detergents) that comes from our households and cities. Together, this runoff provides nutrients which make the water bloom with algae, and the fish are choked to death.

Of course, at the same time we are adding to the same bodies of water the chemicals in acid rain and the industrial toxins that have the effect of killing the algae, as well as everything else in the water. It would be nice to think that we could balance these elements off, so that a single lake could use the two kinds of assaults to nullify each other, the chemicals that cause eutrophication coming from one source, the poisons that kill life coming from another, so that they would neutralize each other and all would be well.

That would be wonderful, but it doesn't happen. Nothing is ever that easy. Even if we could help out the balance by pouring in a little extra of one kind of contaminant or the other, it would not save the biological balance of the water. By then it is too late.

When eutrophication occurs and the algae blooms, all the fish don't die at once. It might be better if they did, because the species of fish that die first are generally the ones we would most like to preserve, largely the usual sports fish.

Illinois is a case in point. In 1989 the state's Environmental Protection Agency reported that 90% of all Illinois lakes (Lake Michigan was not included) were dirty enough to take a lot of the fun out of recreational uses. Fishermen found that clean-water sports fish like bass and pike had been replaced by dirty-water fish like carp and bullheads. Part of this was

due to silting from farm erosion, lumbering and construction projects. Even more was the result of algae blooms that
came from the eutrophication brought about by agricultural
and lawn runoff; as predicted, the algae consumed so much
oxygen that the sports fish suffocated. Eutrophication doesn't
help those who want to swim in the lakes, either, as the environmental expert Toby Frevert (of the Illinois EPA) pointed
out. He searched for words of consolation for those concerned about the recreational uses of his state's waters, and
could find only a few: "You may not enjoy boating as much
on a pea green lake," he said, "but you can still do it."

Scientists tell us that we can't expect any lake to last forever. In the geologically long range, all lakes are temporary
things. It is the natural destiny of any lake to silt up and become a marsh and ultimately dry land: that's what lakes are,
sediment traps in streams and rivers. To try to "preserve" a
lake as a permanent part of a landscape is to attempt to interfere with our old friend Gaia and, in the long run, likely
to be impossible. We can accept that remote future prospect
well enough, but it is the timing that is important to us. The
slow death of a lake over thousands of years is easy to adjust
to, while accelerating the process by fouling them with plant
growth over a matter of decades is not.

Poisoning the life out of them is still worse. If some of
Lake Michigan's fish can no longer be eaten freely, at least
there are still fish there. There are lakes and streams in
Northern New England and eastern Canada that are not so
fortunate; the water is clear and clean, but the fish are dead
from acid rain.

Here too the problem is not just ours. The rest of the world is experiencing the same deaths. You remember from our discussion of air pollution that England had solved most of its smog problem by exporting the smog to Scandinavia and northern Europe, where it falls as acid rain. (Of course, it isn't *all* English airborne pollution—every one of the countries of Western Europe contributes a share.) That acid rain has already killed off so many sport-fishing rivers in Sweden, for instance, that the Swedish government has invested $20,000,000 in a project to restore the chemical balance of the rivers by adding lime to the water.

As technological fixes go, that seems a plausible thing to do. Any high-school chemistry student knows that when you add a basic chemical like lime to an acid they neutralize each other—in the test tube.

In the real world of the poisoned rivers, it does not work as well as in the chemistry laboratory. For one thing, adding another chemical to a body of water in itself alters its ecology. This means that the particular cluster of species of sports fish—and all other living things in the system—that evolved to meet the pre-acidification environment in the natural river are not necessarily the same ones that will thrive in the limed one. Fish can be restocked after liming, but it is unlikely that they can be the *same* kinds of fish.

But even the new species of fish will be at risk, because of the question of snow.

There's plenty of snow in a Swedish winter. When spring comes along, the snow melts and the meltwater goes into the rivers. That's fine in itself, except that that melting snow has been accumulating acid fallout all winter; and if it happens to melt quickly, pouring great gushes of intensely acidified floodwater into the rivers, the acid level climbs rapidly. One sudden warm spell in early spring may poison every restocked fish overnight.

There is also a serious technical problem in deciding *how* the lime should be added. If you pour it into the water itself, then the next heavy rain will dilute the lime—while simultaneously adding new airborne acid. Result, more dead fish. If instead you pour it onto the ground nearby, so that when rain does fall it will obligingly wash a new supply of lime into the river or lake, you have to guess very precisely just how much of the chemical to put where. The guesses are seldom right. Worse still, the lime itself is a chemical poison which will kill off the plants and animals that it touches where you have dumped it on the ground.

As with nearly all technical fixes for our environmental problems, the fix of liming an acid river solves one problem only by creating others. There is only one long-term real fix to the problem of acid rain. That is to keep the pollutants out of the air, and thus to prevent the acid rain from falling in the first place.

We've been talking about the amenities we get from our lakes and rivers, and, of course, amenities are by definition things we could live without if we had to. Let's turn to some of the more basic needs, starting with our requirement for irrigation water for our farms.

It isn't quite true that we *need* irrigation to grow food. Before irrigation was invented farmers still grew crops watered by natural rainfall; over many parts of the world, farmers still do. The problem is that there is a lot more of the human race around now. Without irrigation water we can no longer grow *enough* food to feed our present 5.2 billion human beings. Natural rainfall won't do the trick. Many of the most productive farms in the world are in areas where it simply doesn't rain enough—California's marvelously fertile Central Valley can expect only about five inches a year, less than some actual deserts. Even the best farm areas suffer occasional drought years; in the United States the midwest was hit so hard in the summer drought of 1988 that seedlings shriveled in the caked and cracked soil, and the southeast suffered even more a year or two earlier. If for any reason irrigation could no longer be practiced at least a billion of the world's people would starve.

And yet the sources of water for irrigation are being exhausted or spoiled faster than they are being replenished by nature.

We've already mentioned the Ogallala aquifer, but it is not the only lake of underground water—"fossil water," it's called, because most of it has been there for thousands of years— which is being depleted rapidly, either here or abroad.

The Mexican peninsula of Baja California demonstrates aggravated human agricultural folly. Baja is entirely dependent on fossil water—the peninsula gets no more natural rain than Death Valley. That does not stop the farmers. Water-thirsty crops like cotton, which need the equivalent of a hundred inches of rain a year, are being grown in large quantities on lands which don't get more than four. Obviously that can't go on forever, and in fact the bill is now coming due for those Baja farmers. As the irrigation water is pumped out, sea water is seeping into the aquifer. Now the wells nearest the ocean

are turning brackish, and the Baja farmers are scampering
to seek salt-resistant crop strains to replace the traditional
ones they're now growing.

In the drying of the Aral Sea in the USSR irrigation to grow
cotton is again the culprit. The water of the Aral Sea comes
from two big river systems—the Amul-Darya and the Syr-
Darya—and it is the water-thirsty cotton plantations that
have diverted so much water out of those rivers that the Aral
Sea has retreated from large sections of its shoreline, turning former waters into mud flats.

Like the Californians, the Soviets too looked for new water
supplies in preference to reforming their farm practices, and
the direction they looked in was also north. Before Mikhail
Gorbachev came along, the previous leaders of the USSR had
begun an immense water-diversion project to drain two great
Siberian rivers, the Ob and the Yenisei. The plan was that
much of their water would flow south to replenish the
drying areas of habitation and farming, instead of flowing
"uselessly"—as they described it—northward into the
Arctic Ocean.

But what does "uselessly" mean? What are the natural
"uses" of the outflow of great rivers? We do not always know
how useful a river's outflow may be until we've already
stopped it, as we saw when the Egyptians curtailed the "useless" flow of Nile water into the Mediterranean with their
Aswan Dam and found that by doing so they had exterminated some of their most valuable commercial saltwater
fisheries.

So scientists worldwide began to point out that there were
great dangers in the Ob-Yenisei diversion plan. The eco-
system of the Arctic is still poorly understood, but it is a fact

that those two rivers have contributed vast amounts of fresh water to the northern seas. No one could say for sure what the consequences of the diversion would be, but there was every reason to fear that they would have had unpredictable effects on the climate of the entire northern hemisphere if the plan had been carried through.

Fortunately for all of us, it wasn't. Mikhail Gorbachev came to power in the USSR. Individual citizens—even environmentalists—began to have some say in what their government did. Glasnost and perestroika asserted themselves, and in 1986 the river-diversion plans were canceled . . . but the Aral Sea is still shrinking.

Amenities are pleasant, irrigation is important, but there is one other human need for fresh water that can be described only as urgent:

We need it to drink.

When we talk about the need for drinking water we are speaking about the needs of our cities. Rural areas don't have that problem, because by the time drinking water got scarce the water the crops need would already be long gone, and so would the farmers. America saw that happen with many midwestern farmers in the Dust Bowl of the 1930s in America. It goes on happening in such chronically drought-stricken areas as the Sahel.

The cities, however, have a huge, permanent and growing requirement for supplies of clean, fresh water, whether there's a lot of rainfall that year or not. They can't get along without it. The half-gallon or so a day every human being must drink to live is not the end of it. (It may be more in hot weather—troops in the Persian Gulf, for instance, were urged to drink

six gallons a day . . . all of which had to be carted to them.)
Besides slaking the thirst of their human inhabitants, cities
need water to wash their dishes and spray their lawns and
flush their toilets; so instead of half a gallon a day per person,
the city needs to provide more like eighty gallons a day out
of each citizen's water taps.

The concentrated wealth of cities (even poor ones) lets
them outbid almost anyone else for water supplies. So they
usually manage to find what they need—as long as nature/
Gaia cooperates in giving them any reasonable amount of
rainfall at all.

Sometimes it doesn't. Then there is trouble. There is not
much that even a rich city can do about it when nature simply withholds rain. When the reservoirs run low cars can't
be washed, restaurants are ordered to stop supplying water
to their customers (to the delight of their bartenders) and
other less dainty measures are taken. For instance, there is
former New York mayor Ed Koch's memorable dictum in a
time of local drought in the early 1980s: "If it's brown, flush
it down; if it's yellow, let it mellow."

New York has been on the edge of a water crisis for years,
and the occasional drought in the Catskills (where its main
reservoirs and catchment areas are) is only part of its problem.
Its worst danger of water shortage lies in the New York water
supply system itself.

What is wrong with New York's water system is what's
wrong with much of America's infrastructure—not only
water supply systems, but sewers and for that matter bridges,
transit systems and tunnels as well. Decades of budget cutting have left their mark. The Catskill water comes to the
city through two aging aqueducts, built before most New

Yorkers were born and so corroded through age that the operators no longer dare use the huge old valves that control them. Once the valves are turned off, the operators can't be sure of being able to turn them on again. Within the city water mains burst from time to time, flooding out subways and drowning cellars. But that can be tolerated, while a single accident to a main aqueduct would leave the city parched until a new one could be completed. (The needed third aqueduct has been "under construction"—as the flow of money allowed—for half a century, and is still nowhere near done.) And, of course, every hour of every day there are leaking pipes and fixtures that waste almost as much water as gets used in some places. Other cities are in almost as precarious a position because of leaky and nearly collapsing delivery systems. Some are in worse, and, of course, not only in the United States. A study in England showed that the money a number of the largest water authorities were planning to spend on the construction of new reservoirs would actually produce more water for their customers if it were spent instead on replacing old cast-iron pipes and leaking valves.

As we find over and over again in the study of natural disasters, the "natural" droughts are made a great deal worse, when not actually caused, by human activities. New York's droughts are a source of astonishment to all Westerners and all Europeans. There seems to be no reason for the city ever to run out of water: How in the world, with one of the country's greatest rivers running by its doorstep, can New York City go dry? The answer is that the Hudson's "fresh" water has been terribly staled by all the cities and industries upstream long before it reaches New York City. By then it is not merely unfit to drink; it is too filthy even to swim in.

The whole world suffers from a lack of really clean sources of urban water. Almost every city finds it necessary to purify its water, from whatever source, before its inhabitants can safely drink it.

Water-purification systems are expensive, but they work fairly well. Aeration fountains and chemical treatment can kill off most of the disease organisms with which people have fouled their water supplies, such as bacteria, viruses, helminths and other parasites—at least, they can do the job when the city is rich enough to afford proper treatment. The toxic chemicals that seep into the water sources from industry, agriculture, waste dumps, leaking underground gasoline tanks or the simple practice of dumping old oil into a sewer when a car goes through an oil change (it has been estimated that as much used automotive oil is dumped into American sewer systems every twenty days as was ever spilled by the *Exxon Valdez*) are harder to remove. Sometimes they are nearly impossible.

Still, by and large, cities get the water they need, one way or another. That's not the end of the story, though. Even after a city has managed to bring more or less pure fresh water in for its thirsty people and industries, it has solved only half its problem. The other half is what to do with the water when the city is through using it: the sewage.

Think of a city, any city, as a machine for polluting water. Fresh water goes in at one end of the machine. Sewage comes out at the other.

A few hundred years ago the disposal of that sewage presented no real problem—well, yes; there were *problems*. When householders were in the unattractive habit of dumping their

chamber pots out of upstairs windows into the streets the results were not only unpleasant for the passers-by, they caused
a lot of disease. But no large city does that any more. Once
actual sewers had been invented and installed, the scraps
and excrement could at least be made to vanish from the city
streets.

Sewers, however, can only carry wastes away; they can't
make them disappear. The wastes have to go somewhere.

What, for example, the city of London did with its sewage for many centuries was to dump it back in the River
Thames, trusting that the natural processes of purification
would clean it up enough so that people downstream wouldn't
all sicken and die. So did most American municipalities, and
indeed all of the other cities in the world that had a nearby
river. Seaside cities were sometimes a little more prudent,
because their beaches were an economic asset, but then they
were luckier, too. The ones in America had a whole Atlantic
or Pacific ocean at their front doors: long pipes carried the
sewage out into the open sea, and if now and then some of
it floated back onto the beaches, at least that didn't happen
often.

Growing populations made all that, as so many other things,
impossible. Now, in almost every city in the United States
(not, unfortunately, in all the rest of the world), the wastewater still goes into the same sinks of river or ocean, but generally at least now it is treated to remove some of the worst
components before it is disposed of.

The greatest part of the volume of municipal sewage is
nothing but water. Some of the water comes from storm
drains, which catch and carry away the runoff from rains—
and, in the process, scrub the streets and sidewalks of

whatever trash has accumulated. Storm water can be quite unappetizing, but it is not usually a great danger to health. Most of the rest of the city's sewage outflow is worse. That part is the water, along with the things the water carries, that comes from the kitchen sinks, bathtubs and flush toilets of the people who live and work there. That portion of the sewage is a thin solution of food scraps, soap and detergents, human excrement, toilet paper, sanitary napkins and other organic, and often highly infectious, odds and ends of human waste.

That sounds even more unattractive than the water of storm drains but, really, that part of the sewage problem is easy enough to handle. Well designed water purification plants can make this "reprocessed" water palatable and safe— as hundreds of municipal sanitary engineers have gone to the trouble of proving, for audiences of newspaper reporters, by drinking a tumbler of the effluent.

Typically, a wastewater processing plant pumps the raw sewage into settling tanks. The solids precipitate out, having been helped to do so by various chemical agents. The remaining liquid is treated with air bubbles and chemicals to kill off everything alive, including disease organisms. Then, when it is no longer a significant health hazard, it is simply poured back into the nearest body of water—generally the same body of water it came from.

There still remains the solid sludge that has settled out.

That sludge does not start out really solid. At first it is a thick, stinking soup collected in settling tanks half the size of a football field, holding as much as a million gallons of the stuff. There it is usually treated with bacteria that "digest" the raw material. To make the process work it has to be stirred, which in some modern installations is accomplished by huge built-in vanes that are slowly turned by a powerful motor. (Other installations are more primitive. In some of them some sewage worker has a nasty little job: he climbs onto a kind of outboard motorboat and spends two or three hours circling around the surface of the tank, churning up the sludge.)

When all this is properly done and the remaining solids are dried, they could have commercial value. By then what is left of the sludge no longer has an offensive odor and is not infectious; it could be a valuable mulch or fertilizer.

It isn't often used that way in the real world. The reason is that almost all city water systems are used by industry as well as by households.

The substances that industries dump into their sewers are far more dangerous than simple excrement. They include oils, solvents, acids, metals and toxic wastes of all kinds from manufacturing processes. So in most cases municipal sludge is not clean at all.

The elemental metals that turn up in industrial wastes are bad enough. Zinc and cadmium are the ones most likely to be taken up by living things, and they are both poisonous. Therefore if sludge which contains them is spread on farm-land, as it might otherwise be with good effects, those metals cause trouble.

Zinc, although it is toxic to human beings, seldom poisons human systems directly. The reason for that is that it is even more toxic to plants; therefore, the plants which take it up from, say, irrigation water, die early of poisoning and thus are not eaten by any animal, including ourselves.

On the other hand, plants are pretty good at tolerating

cadmium. Animals are not. So the plants grow well enough, with their content of toxic cadmium and all, and when they are then eaten by animals, and ultimately by human beings, the cadmium accumulates in the kidneys and causes serious illness or even death.

Lead is also common in sewage sludge, and also deadly. Steve McGrath, a scientist at the Rothamsted Experimental Station in England, studied three test farm plots. An early experiment, between 1942 and 1961, had fertilized one with processed sewage sludge, the second with barnyard manure, the third with standard farm fertilizers. In 1990, McGrath found that 85% of the original metals still remained in the sludge-treated plot. When he planted red beets, the sludge-treated plot's yield was only half that of the one that had, so long ago, received only barnyard manure. (Curiously, chemically barnyard manure is very like sewage sludge . . . but without the toxic metals.)

The ones we have just named don't end the list of dangerous metals in your average city sludge: copper and nickel cause damage too, and some evidence shows that chromium (usually from tanneries) is also a risk. But the metals aren't the only problem. Compounds from various kinds of industry are even more deadly: PCBs, dioxins and a host of others can kill even in tiny dilutions, or cause birth defects, or lead to cancers, and there is simply no effective way to get these chemical poisons out of the sludge.

So the sludge is not only useless, it has become a major disposal problem itself. If spread on the ground, it kills plants. If buried, its seepage may pollute underground aquifers. If dumped at sea, it pollutes the sea . . . and that brings us to the other kind of water, the salty stuff that makes up our oceans.

* * *

Most of the Earth's surface is covered with sea water—roughly 300,000,000 cubic miles of the stuff. For anyone who has flown across an ocean it might easily seem that nothing human beings could do could seriously damage all that immense volume of water. In fact, that is exactly how it *did* seem to almost everyone—until scientists began to take stock of what was going on.

Then the picture changed.

Perhaps the most startling and dramatic picture of what has happened to our oceans came from Thor Heyerdahl. Drifting in his papyrus raft, *Ra*, in the empty South Atlantic, far from land or even any shipping lane, Heyerdahl leaned over the side to gaze into the water . . . and found blobs of degenerated oil and other bits of human waste floating even there.

So even the least traveled seas are showing signs of human pollution; but, of course, the worst damage is in the waters closest to land, where the people are. This "continental shelf" water amounts to only one two-hundredth of the great volume of the world ocean, but it is the living part. It is where most of the sea's life is, and it is the part that is most important to the global environment.

It's a pity that it is also the part of the sea where human activities do the most damage. It is there that we dump our sewage, drill our offshore oil wells and suffer most of our supertanker oil spills; there that our polluted rivers discharge their own burden of sewage and their agricultural and industrial wastes. While we add to oceanic pollution in all these ways, we also cut back on the inflow of fresh river water that would dilute it by damming the rivers. The dams that are meant to make hydroelectric power ultimately release their water to the sea, but irrigation dams do not; and by the end of this decade two-thirds of all the river discharges in the world will be controlled by dams.

So we see red tides in Florida and green slime in the Adriatic Sea, killing fish and repelling human bathers; plus toxic chemicals; plus even grosser contaminants. SCUBA divers found shreds of wadded toilet paper on the sea bottom off the Jersey shore and in many other places. Also along the New Jersey coast, in 1987, dolphins began dying in the Atlantic, and when their corpses were examined they turned out to contain high levels of PCBs. (One dolphin's body fat had an astonishing 6800 parts per million of PCBs—the government's limit on PCBs in fish to be eaten is *two* parts per million. More than a half of one per cent of the unfortunate dolphin's blubber was pure lethally toxic chemical.) Catches for California fisheries dropped 90% from 1982 to 1989. And then came the most shocking event of all, when discarded hospital supplies—hypodermics, soiled bandages, specimens of diseased blood, all the waste of operating rooms—began washing up on American beaches in 1987 and 1988.

It would take this whole book, and several more like it, to outline all the ways in which the world's saltwater seas have been polluted and damaged, from the filthy eastern Baltic, through the chemical-foamed Mediterranean, even to the shores of the Pacific. Even to Australia, for those pretty TV tourism commercials that show the sundrenched beaches near Sydney do not show how raw sewage has made some of those beaches unsafe for swimmers for two or three days of the week. Sydney's sewage is particularly foul—almost 40% of it is industrial waste, much of it toxic—and when there are storms as much as a third of it overflows the treatment plants and goes directly into the sea. Even Scandinavia, world leader in environmental action, has its problems: in 1988 the Norwegian coast was inundated with a yellow-brown algal slime, and the Norwegian Navy had to be called in to tow the salmon-farming pens to cleaner water. Even the Netherlands: when you flush a toilet in the Hague, the flush goes through pipes to emerge, untreated, less than a mile away, coming out as a foul chocolate-colored mix at the beach resort of Scheveningen.

Our destructive pollution of the oceans comes in a thousand forms, in tens of thousands of places. Around islands in the Caribbean the coral reefs are dying, choked to death from silt caused by the deforestation of the islands' forests. Sewage sludge was dumped far out at sea twenty or more years ago, but not far enough; now it is oozing toward bathing beaches near many American cities. The salt water of New York Harbor has been found to contain the dioxin 2,3,7,8-TCDD— one of the deadliest poisons known, about a hundred thousand times as lethal as DDT. There is a five-ton container of the violently toxic pesticide lindane—so dangerous that its use is banned entirely in seventy countries, and it is manufactured now only in two places in France and Spain—lying at the bottom of the English Channel. (Dr. Paul Johnson of Greenpeace says, "If the lindane gets into the sea it could kill all marine life in a strip of sea 440 miles long, and would last for twenty years.") There are even dangerous nuclear materials as well—fallout from French and American bomb tests in the Pacific, nuclear wastes dumped in the Atlantic for lack of any better ideas as to how to dispose of them—or at least to get them out of sight. Fifty nuclear warheads and nine nuclear reactors lie on the sea bottom in places all over the world, the result of submarine and other ship sinkings and accidents; what will happen when their casings corrode and their radioactive contents are released into the water?

At least most of those nuclear reactors do have casings. Imagine what would happen with a Chernobyl-style accident at sea.

One such very nearly did occur in the northwest Atlantic, on July 4, 1961. A former crewman of a Soviet submarine has gone public with the story: A reactor overheated, there was an explosion. Eight of the crew were killed outright, and a fire broke out. Somehow the survivors managed to bring the submarine to the surface and got the fire out before the fuel melted . . . barely.

And then there are the oil spills.

They come from tankers, pipelines and offshore oil wells. They come in great volume. For a quarter of a century oil tankers have been running aground, catching fire, colliding or breaking up in storms. The oil they spill poisons fish, chokes bottom organisms, fouls beaches, kills seabirds and seals . . . and produces the nasty lumps of oil-residues Thor Heyerdahl saw in the middle of the sea. Not all the oil comes from accidents. A great deal comes from tanker captains rinsing out their tanks with seawater, far from land. . . .

But there is no way, any more, for them to get far enough. Now the whole sea is at risk.

Can we calculate an SAP for the pollution of the world's water?

Yes, for some things. With careful control of fertilizers and sewage, the kinds of pollution that produce algae green lakes and red tides can be limited to the levels the water can cope with. The presence of toxic chemicals in drinking water could be limited, too, though that would necessarily have to be kept to much lower levels.

The pollution of the oceans with heavy metals and radioactive materials, however, is harder to calculate. What we hope to do with most pollutants is to take them out of the parts of the environment where they do harm and remove them to a harmless "sink." But the ocean itself is the ultimate sink. There is no other place for poisons to go.

This chapter does not end the catalogue of human damage to our bodies of water.

For example, we haven't even touched on the destruction of marine life through non-poisonous means. Plastics are killing fish in vast numbers: membranes from six-packs; "ghost nets"; Saran wrap. Sea birds are strangled by them, turtles, thinking the transparent plastics are the jellyfish they feed on, eat them and die, fish are trapped in lost nets and die there—and those miles-long trawler nets, as durable as any good plastic, go on killing marine life for decades after some fishing vessel has lost them. And all this is in addition to the esthetics of—for example—picking beer cans off a coral reef or finding plastic Coke bottles washed up on the beach at Big Sur. . . .

But we are beginning to trespass on the subject of the ways in which we trash the world in general, and that's better dealt with in the next chapter.

11

Trashing the World

Besides the life-threatening crimes against the environment we've already talked about, we're charged with a long list of smaller damages to the planet Earth . . . such things as damaging the soil itself, the food we eat, our own health and the survival of the other living things, plants and animals—something like 30,000,000 species of them—who are our fellow passengers on Spaceship Earth.

They all matter. It would be wrong not to mention them. But there are so *many* of them, so let's hurry through them as best we can in this one chapter.

The Soil

Soil is the thin crust on the surface of the land that plants can grow on, and we need it to live. Lands which lack soil we call deserts. Deserts come in two varieties: the ones that are natural, like the Gobi in Asia, and those, like much of the Sahara, which we (or our remote ancestors) have created ourselves.

It's not difficult to turn fertile land into desert. Doing that is simply returning it to what the whole land surface of the

Earth looked like a few billion years ago, for every last bit of soil on the planet started its existence as naked rock.

We see that process still going on in volcanic islands, like Hawaii, where from time to time Kilauea erupts and buries another housing development—or stretch of farmland—under a river of lava. When the lava cools it is bare rock. Nothing will grow there. As time goes on the rock gets chipped or cracked—by waves at the water's edge, by wind, by rainfall—until there are tiny crevices where dust can accumulate, and a few airborne seeds can begin to sprout. Then the roots of the growing plants themselves help to break up the rock as they swell and press against the crevices. When the plants die their remains enrich the newly formed soil. Insects and worms arrive to break it down still further, and to aerate it with their burrows, and, like the plants, to fertilize it with their droppings and their remains. In a few years the lava flow has begun to show green; in a few decades, it begins to be a tropical forest; in a few centuries, it has become one of the richest of all soils.

That is the life story of soil everywhere. It is proper to call it a "life" story, for soil is really alive. It is continually replenished by the tiny organisms it contains. And, since it is alive, it can be killed.

Most frequently, the killers of soil are ourselves. Sometimes we do it simply by removing the covering of vegetation that protects it, say by clear-cutting forests; we did it long ago in the United States when farmers first began plowing up the native buffalo grass to plant crops. While the soil was bare it was at risk; if rains did not come in time it dried up and blew away: thus the great Dust Bowl of the 1930s. Although American farmers have learned something from that experience, they have not stopped the process. American farmers still lose millions of tons of soil through erosion, which can remove an inch of exposed topsoil in a single heavy rain.

Here again there is a race between creation and destruction. Nature—or Gaia—has not stopped making new soil for us, even in drought-stricken areas. But to restore that one lost inch, natural processes may take a century or more, and every time a midwestern farmer ships a ton of wheat to an overseas customer he also ships a ton of soil to silt up on the delta of the Mississippi River through soil erosion.

As in so many other ways, here too the United States started out more blessed than most of the world. The settlers of the American midwest found soil so deep that, even after many decades of abuse, substantial amounts remain. Tropical soils are much thinner—despite the luxuriant appearance of an Amazon rain forest, the soil is only inches deep—and tropical farmers found long ago that when they cut down the trees to plant their farms, the exposed soil supports only a few crops. After that it is worn out, too depleted to produce any useful yield (and then the farmers move on to cut down another tract of forest).

In other parts of the world the local people denude their soil for other reasons as well, often by rooting up every shrub they can find for fuelwood. Many Asian hillsides are now bare for this reason.

We'll have more to say about deforestation in a moment, but it doesn't matter why or how we do it; once we kill off the plants, whether they are giant redwoods or buffalo grass, the soil itself begins to die.

Sometimes we kill the soil deliberately, particularly in warfare. In the Vietnam war we did a lot of that through the use of defoliants like the celebrated Agent Orange. This was particularly thoroughgoing destruction, because much of Vietnam's farmland is what is called "laterite soil." Once laterite ground is cleared of vegetation the soil changes its character. It becomes hard, insoluble and almost useless for growing anything again. (The "stone" that the famous Cambodian ruins of Angkor Wat are built of is laterite, and those carvings have endured for centuries.)

But all soils are vulnerable through erosion, once the ground cover is removed. This race, too, is being lost.

Removing soil bodily through erosion is not the only way in which we damage the earth we grow our crops in. Even when the soil is still there and looks as healthy as ever, our farming habits may still have left it depleted in the nitrogen, minerals and other important constituents that the crops need to grow.

When that happens the farmer's crop yields suffer. He can't make his mortgage payments without full crops, so he has to try to get the yields up again. His usual first choice is to sow the land with artificial fertilizers.

Nearly all American farmers invest heavily in chemical fertilizers as a matter of course, in spite of the fact that there are better, more natural ways of dealing with, say, the loss of nitrogen. A wheat farmer could accomplish the same thing by rotating an occasional crop of legumes between wheat plantings, for the legumes (with the aid of the kind of symbiotic organisms in their roots called mycorrhizae) have the capacity to *add* nitrogen to the soil. Many farmers know this and would like to do it, but sometimes they aren't allowed to—because government regulations are so structured that their agricultural support payments may be taken away from them if they switch to a different crop.

Those same support payments have an even worse effect. Although there is a large (and expensive) "soil bank" program, those same subsidies encourage farmers to plant every other acre of soil that can be teased into producing a crop—even to produce a crop we don't need, one which is in oversupply and has no commercial market. The result is that thin soils, soils on slopes, soils where water is scarce and irrigation is hard-pressed and soils of many other kinds that should never be farmed at all are being forced into service.

Perhaps that could be forgiven if it were an emergency short-time measure designed to feed starving people . . . but should it be allowed when its major purpose is to earn those agricultural subsidies?

Deforestation

When we spoke of tropical farmers cutting down rain forests to plant their crops, it may have sounded like the typical human short-sightedness that is at the root of so many of our environmental problems.

Actually, it wasn't, at least not before big business got into the act. Actually, those pre-industrial farmers were smarter than they seemed. They cut only small tracts of forest at a time. When they moved on, the clearing they left was surrounded by intact forest, and the trees and plants around the clearing reseeded the stripped tracts quite quickly; within a relatively short time they were almost indistinguishable from the virgin forest around them.

At the same time, those people valued the intact forest itself; it supplied them with many of the things they needed, without the necessity of planting crops. If that is changing—if the rain forests are disappearing, as they are—it is not because of the stupidity of the natives. It is because of the greed of large-scale exploiters, who want to cut them down so they can sell the valuable tropical woods—to us—and are even willing to kill to do it.

Often enough, the native people know better. They can look at the rain forest as a sort of government bond. If they hold it, it will produce perhaps a thousand dollars' worth of benefits a year, every year, forever. If it is cut down—for the sake of the lumber, or to clear it and use the land to pasture beef cattle for sale to America, or if it is mined and so destroyed—then it will produce a one-time windfall that is worth (let's say) as much as $50,000 in a short time. And after that it will be worthless.

So there is an on-going struggle here, between the people who live in the world's rain forests and those outside exploiters who want to cut them down. The people who live there value that permanent, annual income that they can count on—from such activities as rubber-tapping, or harvesting nuts for sale, or selectively felling a few trees, as well as the things they get for their own use. This annual "dividend" provides the support for their families and themselves and can go on doing so forever.

But the exploiters look at the forests in a different way. They can bring in the chain saws and make the fast $50,000. Then they can go on to cut down another tract for another $50,000, and keep on doing that until the rain forest runs out—which will be, worldwide, perhaps not much more than another few decades.

The exploiters have everything on their side. They even have guns, as a famous recent case showed. That involved a man named Chico Mendez. He started out as an illiterate rubber tapper in the Brazilian village of Xapuri, and transformed himself into a leading and articulate crusader for saving the forest. He led many of his people in a fight to prevent the destruction of mass logging—too successfully for his own good. Just before Christmas of 1988 he was gunned down.

Mendez knew the risk he was running. Everybody did. As the local prosecuting attorney said, "The fact is, everybody knew Chico Mendez was going to be killed sooner or later."

What is happening in Brazil is happening to all the world's tropical forests, in much the same way: they are being stripped to bare ground, to produce quick profits. Thailand devastated four-fifths of its rich teak and other forests to earn foreign exchange, turning vast stretches of once lovely woodland into a dust bowl. Now, with their own resources almost gone, the Thai lumber companies were forbidden to do any more logging at all. So they turned to the forests of Thailand's neighbor, Myanmar (the country until recently called Burma). Myanmar's new revolutionary government is even hungrier for hard currency, and so a million acres of forest are now gone there, too, with every acre of the rest threatened.

It is easy for Americans to feel indignation at the ecological sins of these Third World peoples—until we realize that all of it is done to produce goods for sale to the rich nations of the world, including ourselves. Or until we take note of the fact that our own enlightened countries are doing quite as badly with our own forests. The great temperate forests of the Pacific Northwest are being slaughtered. The lands owned by the logging companies go first, but even the lands owned by the federal government are at risk. Logging companies can secure the right to fell trees on federal land for a pittance—as one environmentalist put it, a logger can buy a 200-foot thousand-year-old redwood for about the price of a Big Mac.

When the first Europeans visited North America they described it as a mass of impenetrable forests. No more. Most of them had been cut down by the time of the Civil War; most of the rest have gone since. Early in this century, about the only virgin forests remaining in North America, apart from Alaska, were in the Pacific Northwest and Canada. By now 60% of Canadian forests are gone, as are 90% of those in the U.S., and the rest are going fast.

Of course, in much of the world deforestation is ancient history, though its results are still with us—for example, the bare hills of Greece and other countries in the region, their trees long ago cut down for shipbuilding and fuel. Or, if you want to know the ultimate consequences of deforestation, you can do no better than to look at that lonely little outpost of human life called Easter Island in the Pacific Ocean, famous for its strange, great stone images.

Easter Island was well forested when the first Polynesians settled it. It is forested no longer. After a number of generations had kept on cutting down the trees for fuel, boats, dwellings and weapons (assisted by the practice of those other human habits of intertribal wars and strife of many kinds), Easter Island was so bare that the remaining inhabitants could no longer find wood enough to build the canoes that would take them off the destroyed land; and there they all died. By the time the island was rediscovered not one of those ancient early settlers still survived.

If we haven't done as badly as the extinct Easter Islanders, it is mainly because we have had more trees to destroy in the first place. We cannot claim to be more prudent.

As in the case of the Brazilians, we cannot even claim that we are making ourselves richer by what we do—not if we take the whole economy into consideration—for many of the dollar profits that come from cutting down a forest are matched by the dollar losses somewhere else in the economy. For example, in Idaho in the 1960s, lumber companies harvested forests along the Salmon River. They made a profit of $14 million for the lumber, but the silt that ran off from the eroded land they left cost the river's salmon fisheries more than $100 million in damages.

Nor have American workers gained from all the mad rush to clear off the forests. Unemployment is heavy in that area, and most of the jobs that have been lost are due less to conservation measures than to the new and widespread practice of allowing export of raw logs to lumber mills overseas. The jobs go with them. Most of the countries of the world have passed laws to forbid such raw-log exports, so their own workers can do the processing of the lumber. The United States has not been that wise.

All in all, we lose 25 million acres of forest every year. Very few of the felled trees are replaced.

Soil Destruction by Urban Sprawl

One of our most efficient ways of destroying American cropland is simply to build on it. That isn't an American monopoly—Egypt loses more cropland to building each year, as the city of Cairo sprawls ever wider, than it gains by increased irrigation from the High Dam at Aswan, and al-

most every city in the world is expanding in the same way—
but we Americans are particularly good at it. Every year we
sacrifice about 800,000 acres of farmland to urban sprawl.

The way cities begin makes that inevitable. Cities are ordinarily founded in the places where the first inhabitants can
best grow crops to survive—that's why the first settlers
choose them. Then the process of city growth continues and
accelerates through the building of highways and airports,
suburban shopping malls and their accompanying immense
parking lots, as well as the housing developments that eat
up countryside. The urban sprawl takes over: When New
York City spread onto Long Island, just after World War II,
the huge Levittown development covered some of the best
potato-growing soil in the world—and was constructed in
such a hurry that that year's potato crop was left to rot un-
harvested, buried in the ground under the new houses.

Unfortunately, construction isn't always that fast, and that
fact also has bad results for the soil.

When the bulldozers come in to begin a large construction project, say the beginning of a 500-unit new condo development, their first task is usually to knock down any trees
on the land; trees get in the way of the construction machinery. The next step is to grade the land, by filling in all the
little ravines and smoothing off the hillocks, so that the site
will be as conveniently level as possible.

This would mean burying much of the site's topsoil, which
most contractors are wise enough to be reluctant to do. So
they make an effort to save it. They carefully scoop up all
the topsoil and pile it in a hill at one corner of the property,
meaning to replace it over the basic substrate once grading
and building is complete.

But building does take time. There are unplanned delays: the builder has to wait for inspectors and permits; he may run out of money and abandon it for a while; almost always, he will put up a few units and postpone the rest until he sells enough to finance the completion of the project. Meanwhile the scooped-up topsoil sits there, heaped up and without vegetation to cover it. The soil can't maintain its vitality under such conditions. It dies. Rains leach out the organic matter, and when at last the final units are built and the soil is carted back to make lawns it isn't topsoil any more. Perhaps it has not yet died completely, but it is certainly no longer what it once was.

The result is that when the homeowner moves in he has to coax it to grow the grass and shrubs he wants to complete his landscaping by feeding it with mulch and heavy doses of chemical fertilizers.

(The destruction of topsoil is not the only folly we commit in our building habits. Why do we build on unstable soils in earthquake areas, for instance in the Marina district that was so badly damaged in the 1989 San Francisco quake? Why do we build in flood plains, when we know that sooner or later all those cellars will be filled with floodwater? Or on barrier islands in the Atlantic, when we know that sooner or later another Hurricane Hugo can sweep all those structures away? Or on hillsides around Los Angeles, when the slopes are known to be unstable and houses slip downhill to destruction in heavy rainstorms?)

Nor is home construction the only industry that destroys soil. Open-cast mining is notorious for the ravaged landscapes it leaves behind, in western Pennsylvania, in Alabama, in the Appalachians, in Wyoming, anywhere where the earth is scooped away to get at the coal or ores beneath; those huge and hideous scars are contaminated with acids created when the ore is exposed to air, and can never be made whole again. Stack pollutants from coal-burning power plants and industries have devastated areas nearby by poisoning the soil with their fallout. (It is not only air quality and remote lakes that suffer from smokestack industries.)

Nor does industry have to squat on top of soil to kill it. The poisons that some industries leach out do the job. Paper mills, for example, are generous with their poisons, particularly when the paper they make is white. To make paper white requires an extra step: before the paper is dried the pulp is bleached with chlorine compounds. In the process some of the chlorine turns into the chemical poisons called dioxins, in particular one, with the long chemical name of 2,3,7,8–tet rachlorodibenzop-dioxin—call it just TCDD—has the proud boast of being termed the most toxic substance ever created by man. This dioxin has been shown to cause damage to living things in the tiniest concentrations the most sensitive instruments can measure, down to a few parts in a trillion. It is bad news, and when paper is bleached with chlorine this TCDD is known to turn up in the mill wastes. It poisons the land nearby, as well as the water and the air; it even turns up in the finished products the paper is made into. When these finished products are containers for food, it even turns up in the food they contain. (Canada had a recent scare when it was found that milk sold in paper cartons made by this bleaching process contained traces of this TCDD, with what effects on the health of the children who drank it no one yet knows.)

The great reality of Gaia, you see, is that all our destructive practices are interconnected. Airborne toxins fall into waters and soak into soils; soil poisons are blown as dust back into the air. If we produce dangerous chemicals they find their way to some living thing sooner or later, by any number of routes.

Sanitary Landfills

Sanitary landfills—less poetically called "garbage dumps"—are supposed to deal with harmful, or merely undesirable, waste. For most of this century almost every American city has trucked its refuse to some isolated tract, sometimes a swamp (thousands of acres of productive wetland have been killed in this way), sometimes merely a stretch of vacant land no one has yet thought to build on.

A typical sanitary landfill is a smelly mesa of decaying organic matter and unpleasant industrial wastes, inhabited by rats and other vermin. It is not at all the sort of thing you want in your backyard. No one else wants them there, either, and they have become one of the critical environmental problems facing America today. Almost all the existing landfills are reaching their maximum capacity, and we are running out of places to put new ones.

The worst of these unsanitary "sanitary landfills," however, aren't municipal. They are run by our own government, especially its military branches.

The Pentagon runs the biggest business establishment in the world. In the process it produces an immense amount of waste of all kinds, from high-level radionuclides to ordinary garbage, from vehicle and aircraft exhausts to toxic chemicals—perhaps half a million tons each year of what is legally defined as hazardous waste. Few of its establishments

comply with either federal or state regulations on its disposal.
Not even when it is dangerously radioactive.

The seventeen American factories that produce nuclear
weapons or their components have deliberately released radioactive iodine into the air and dumped radioactive chemicals in open pits, where wild animals used them for salt licks
(Hanford, Washington), dumped hundreds of thousands of
tons of uranium compounds into streams (Fernald, Ohio),
leaked radioactive wastes into the ground (Savannah River,
South Carolina—which was sited above the Tuscaloosa
aquifer, and has now been discovered to lie directly above an
earthquake fault), killed thousands of sheep by releasing
nerve gas into the atmosphere (Dugway, Utah) and dumped
toxic chemicals into the water supply of civilian communities (Rocky Flats, Colorado—which has the distinction of
being called the most polluted square mile on Earth). Some
of the dangerous "waste" produced by military activities is
deliberate. For instance, the A-10 attack aircraft uses depleted
uranium in its tank-busting cannon shells because the uranium is very heavy and gives them extra mass for penetrating
armor. On impact the uranium burns. It releases micrometer-
sized particles of uranium oxide, which then are airborne
for long distances. Inhaling the particles will release alpha
particles in the lungs, which are carcinogenic.

Nearly all the most dangerous radioactive waste in the
United States comes from military reactors. Some of it,
like plutonium, will menace life for thousands of generations
to come—many times longer than all of human history to
date—since the half-life of plutonium is more than twenty-
five thousand years.

There are 15,000 dangerous military dump sites in the

United States. The cost of cleaning them up—if that is even possible—will run to more than one hundred billion dollars. And we may not know them all.

Some may still be hidden behind the sort of "military security" that conceals nothing from any possible enemy, but prevents citizens from learning about the military's sins. No outsider knows how many accidents involving nuclear weapons have occurred in even the relatively candid American establishment; when such accidents have occurred, nearly half of them were concealed or were officially described as being non-nuclear. This is official policy; *The Bulletin of the Atomic Scientists* has procured a copy of a 1984 navy directive which *orders* commanders to lie to the public in this way when nuclear accidents occur.

This vice isn't just American. The consequences of the 159 nuclear tests that France has conducted in the Pacific were kept secret for years, and it is only recently that the world learned that these tests left, for instance, the island of Mururoa littered with nuclear wastes, some of which—several dozen pounds of plutonium among them—were washed into the sea in a violent 1981 storm. While the Soviets have had a number of military nuclear accidents, the worst of which, as we have seen, was in 1957 in Kyshtym, in the Ural Mountains.

Is there any chance of an American Kyshtym?

There's certainly a suspiciously Kyshtym-like situation at the Hanford Nuclear Reservation in Washington state. Just like the Soviets, the American nuclear-weapon makers took many shortcuts in their work, most of all in the disposal of hazardous, and possibly explosive, nuclear wastes. The Hanford waste tank designated 101-SY contains about a million gallons of mixed wastes, toxic chemicals and radionuclides like cesium-137 and strontium-90 thrown together. Over the years, some of the chemicals have reacted with others. Explosive hydrogen gas has been released. The gas can't easily escape, because a thick layer of crust has formed over the deadly soup, but an observer can see the crust shuddering and lifting as much as a foot while the tank "belches," like the top of an immense pie baking in an oven. If it does escape, and there is a spark of any kind, it will certainly burn, and may explode. How big the explosion might be depends on the largely unknown chemicals that have formed under the crust.

Hanford has already had three "major steam explosions" in such waste tanks. One of them knocked a tank off its foundation and sent radioactive steam boiling fifty feet into the air. In April, 1990, an investigative commission from the Department of Energy surveyed Tank 101-SY. They reported that a Kyshtym-style explosion was not likely . . . but also not impossible.

Hanford has other undesirable distinctions, one of them— its "PUREX" (for Plutonium-Uranium Extraction) plant— is so remarkably dirty that it has failed to meet environmental standards even when closed down. And at least one Hanford event is grotesque.

When in the 1950s the Atomic Energy Commission decided to investigate the dangers of radioactive fallout they systematically exposed 828 beagle puppies over a period of ten years to radioactivity, by feeding or injecting them with radionuclides, or exposing them to radiation. The last one died in 1986, and the experiment was terminated. The dead dogs remained. So in 1990, the remaining "radioactive waste"— which is to say, the bodies of the dogs plus the nearly twenty tons of excrement they had produced in their lives—were buried in 55-gallon drums at Hanford. The burial cost some $26,570 per dead dog.

We can't leave the topic of trashing the world without looking at the ways in which human habits are affecting living things themselves—including ourselves, and the food we eat.

There is certainly a lot of food in the world these days, far more than ever before in history. Agrobiologists have devised some wonderfully productive new crop plants, especially in what was called the Green Revolution, the development of highly productive new strains of rice and other grains in the 60s. As of this writing these new crops and other increases in production have meant that there's even enough food coming off the world's farms to feed every last human of the 5,200,000,000 alive with a decent diet. (There are still famines, of course, but not because the food isn't being grown. Now people starve because of wars, failures of distribution—and, most of all, from simple poverty. But that's a whole other story.)

Yet there are drawbacks to even the Green Revolution. The new strains of grain produce handsomely, but only if they are handsomely supplied with more than usual water, more than usual fertilizers and more than usual pesticides. The costs of raising crops of these miracle grains put them far outside the financial reach of the ordinary peasant farmer of Asia, Africa or Latin America; after twenty years of the Green Revolution it appears that its principal effects have been to make rich farmers richer but poor ones poorer still.

Worse still, the Green Revolution seems to have just about run its course. It added so much edible produce to the world's larders in the first decade or so that the supply of food per capita in the world reached the highest point ever; but since about 1980 the annual increases have not kept pace with increasing population. There is less food per capita in the world now than there was ten years ago, and there is little reason to hope that food production will pick up speed again. Indeed, the prospects are that it may fall even farther behind population as many areas now producing food may have to limit their production or go out of production entirely. The reason for this is that salinization of wells, depletion of water supplies, erosion of the soil, weather, and all the other factors we have been talking about in this book are taking their toll.

There is also an increasing risk of total crop failures, brought about by our custom of growing "monocultures."

When a whole countryside devotes itself to a single "race" of a particular crop—in the farming practice called "monoculture"—no matter how efficient that race is at growing edible produce, it is also at risk. In monoculture, when a single new insect pest, blight or disease appears it may wipe out not just a single farm but a whole growing area—as has happened with some U.S. grain crops in the past. Worse still, when individual farmers give up their traditional crops for new designer strains—as we are encouraging them to do all over the Third World—we lose those ancestral strains which may carry genes we will desperately need for some future crossing.

It is not only the food we get from our farms that may experience catastrophic failures. The large quantity we get from the sea is also at risk.

The damage we do to our world's supply of water, both fresh and salt, naturally damages the fish and shellfish that live in it, as well. But we do more than that. Overfishing has effectively eliminated many food-fish supplies. New England's fishing industry was once the wonder of the world, but it isn't any more. Catches of such staples as cod, yellowtail flounder, haddock and whitefish are down from 50% to 90%, even though there are fewer fishermen to go after them each year. The traditional fish-and-chips shops of London are now pizza joints, partly because the species of cheap fish that supplied them have been fished to extinction. Bottom-dragging trawlers bring up vast quantities of edible fish at one sweep— and at the same time their chains scrape away two or three inches of the bottom, scouring it clean of life. And our habit of drying up the shorelands of the world by diking, draining and building on them has depleted the nurseries where most of the world's fish begin their lives.

And yet, having said all this, we must also say that we do live in a world of plenty—right now.

In America and the developed countries in general, we can see great improvements over even the quite recent past. Our grandparents would not have dreamed of the fresh or frozen products available all the year round in our supermarkets, flown in from faraway farms and seas.

Still, we might question the quality of some of the food so temptingly displayed. "Shippable" fruits and vegetables are usually picked unripe—the taste is different, and the nutrient value is often less. And most of the vegetables in the produce section of the supermarket are grown with intensive applications of fertilizers, pesticides and any number of other chemicals, some of them definitely known to be carcinogenic or otherwise harmful to the health of the consumer. Washing or peeling removes some of them. It may not remove all.

Even the crop chemicals which do not contain anything

that will do you direct harm may still fail to do you much good. When nitrogen fertilizers are applied to food crops the plants grow bigger, but they do not contain correspondingly more nutrients. With heavily fertilized vegetables on her plate, your grandchild may be eating as much of her spinach as ever, but getting less of the things you make her eat the spinach for.

That the pesticides that farmers spray on their crops to protect them can harm human health isn't news, but still it may be surprising to know that they may not even produce any real bottom-line benefit for the farmers. According to a 1990 study conducted by Robert Metcalf, of the University of Illinois, in spite of a great increase in the use of insecticides since World War II, "annual crop losses to insects, which were 7% in the 1940s, rose to 13% in the 1980s."

The worst news of all is that we could even now run out of food, for there is less of it in the world than most people think.

The dismaying fact is that all the world's granaries, supermarkets and household larders together hold only about a 90-day supply of food. That is the slim human margin between plenty and hunger. If at some future time world agriculture should be seriously damaged through warmup or ultraviolet-C, or just through a number of severe droughts and insect infestations striking at the same time, starvation on a scale never before seen could begin in three months.

Shall we talk about the adverse effects of our follies on health and medicine? We know that the news has been getting better for some time; the general worldwide health of the human race is better than any previous century has experienced.

We have wiped out many diseases in the developed world, and a fair number throughout the globe: smallpox, for instance, simply does not exist anywhere on Earth any more, and the crippling and killing effects of polio, tuberculosis and many other scourges have been drastically reduced. Life expectancy has doubled in recent generations.

Improvements are coming more slowly now, though, and some statistics are getting worse. We've used such insecticides as DDT so profligately that many kinds of insects have developed immunity to them. In spite of everything we've done in combating, for example, malaria, there are now over a million malaria deaths each year among African children under the age of five; altogether, according to Robert Gwadz of the National Institute on Allergy and Infectious Diseases, "Insect vector diseases are killing more people today than ever."

So are food-borne diseases, as the very mass-production and distribution measures that have broadened and increased our diet have increased the risk of disease.

For instance, salmonella is flourishing. It has turned up in milk—a few years ago a midwestern supermarket chain experienced a problem in its milk pasteurization which led to hundreds of cases of salmonella poisoning and some deaths. Above all, it turns up in poultry, especially battery-grown chickens. Most raw chickens in the United States are now suspected of containing salmonella organisms. The organisms are very capable of flourishing in the right conditions, so much so that restaurant patrons have gone deathly ill simply from eating a salad prepared on the same cutting board where raw chickens had been cut up earlier. In England, an outbreak of salmonellosis was traced to eggs from infected chickens—the shell was no defense, for the organism had penetrated into the eggs themselves. In the United States, there are 2.5 million cases of salmonella poisoning each year, a sizeable fraction of them also blamed on contaminated eggs, and every year some 2,000 of them die of it. In 1990 the Surgeon General was compelled to issue a flat warning against the consumption of raw or undercooked eggs, but the contamination of the poultry meat itself goes on: the Department of Agriculture, in its tests of processors and distributors, routinely finds that from a third to a half of chicken carcasses carry the disease organisms. Nor is salmonella the only villain. Such other pathogens as *Campylobacter jejuni* are about as prevalent and may be just about as dangerous, though far less well publicized.

To deal with such mass-produced infections, poultry and stock raisers often lace their feed with antibiotics. But this too carries grave risks, for indiscriminate exposure to antibiotics leads to the development of antibiotic-resistant strains; then when someone is really sick, the antibiotic may no longer help him.

Maybe we need another Upton Sinclair, because some of the shocking conditions he described in his novel, *The Jungle*, more than half a century ago can still be found. Animals with ulcers or even cancers are passed for food; the diseased parts are simply trimmed off, providing there aren't too *many* of them. The crowding of the animals in the slaughterhouses (particularly poultry, jammed together and eating each other's excrement) and the filthy waters used to wash, chill and scald the carcasses would turn anyone's stomach.

It is beyond the scope of this book to talk about all of the grave problems in public health today. Instead, let's move on to the things we are doing that affect the thirty million other species we share the Earth with.

For many of them it isn't health but survival that is at stake, and many of them have lost already. We have many ways of exterminating our neighbors on Earth. Hunting wiped out the passenger pigeon and almost did the bison; hunting for ivory or for sport has cut the population of African elephants in half, from 1,200,000 to 600,000 in ten years. Habitat destruction is even more effective; that is what is now impoverishing the flora of the Amazon basin. We kill a great many quite harmless insects with our insect sprays and our electronic bug zappers—thus interfering with the work they do in pollinating our plants; thus killing off some of the plants. The same sprays get into the food chain of birds, and they can no longer lay viable eggs.

Birds are at particular risk. Some species are rapidly approaching extinction. Of the 332 known species of parrots in the world, 329 are endangered; at least seventy of those will probably be gone in the next decade. The lower-48 United States once had a common parrot of its own, the Carolina parakeet; it flocked by the thousands from the Carolinas to Florida, but the last surviving Carolina parakeet died in a zoo in 1918. And for some reason the world's frog population seems to be dying off. There are three or four thousand species of frogs, salamanders, toads and amphibians in general, and, though a handful are holding their own, the great bulk of them are dwindling fast. Their nemesis is clearly man. Because of their biology they are particularly sensitive to pollution of air and water; they also respond significantly to ultraviolet radiation, to temperature changes and especially to loss of habitat through clearing and draining and building. They are, in short, vulnerable to everything that human beings are, only more so . . . and so scientists keep an especially watchful eye on frogs because they represent a kind of distant early warning of what may be in store for ourselves.

Wherever we go and whatever we do, we spread death to other species. When we are children we shriek in pleasant horror as we watch scary movies about killer dinosaurs and carnivorous saber-toothed cats . . . but there has never been a greater killer than mankind.

Not every other species has been harmed by human intervention. Some have flourished, even to the point where they themselves are menaces. We introduce alien species where they wipe out local ones: lampreys in the Great Lakes, rats, goats and sheep on islands, African bees and Asian mosquitoes in the Southwest, kudzu in the South, the "opossum shrimp" in Lake Tahoe, the poison toad, the "stink tree" and the water hyacinth in Florida. There is no end to the list of undesirable exotics that have been imported into the United States: starlings, Japanese honeysuckle, bachelor's buttons, the zebra mussel and both marijuana and the opium poppy. Other imports are not offensive in themselves but do harm in indirect ways—for instance, the imported barberry, an innocuous and even attractive plant which unfortunately serves as host for the stem rust disease that devastates wheatfields.

In the rest of the world, England now has a problem with a little helminth popularly called "the killer flatworm." The worms were accidentally imported from New Zealand; they eat the local earthworms—bad news for English farmers, since it is those native earthworms which aerate the soil and make it more fertile with their droppings. English fishermen also have a problem with the zander, or pike perch, brought to stock British rivers in the 1960s and now voraciously devouring the native sport fish. Africans living around Lake Victoria imported the Nile perch for better fishing; it too ate its way through the lake, so that a third of the native fish species have gone extinct in forty years. In Australia sportsmen brought in the rabbit and the fox—now the rabbit eats scarce grass the herdsmen would like to have for their flocks, and the fox has preyed on native marsupials until some of them are on the brink of extinction.

The most recently settled lands are where we see the clearest picture of the devastation, for instance the island paradise of Hawaii. Any visitor knows that it is filled with beautiful plants and natural marvels, but present visitors are not likely to be aware that very few of the living things that make up Hawaii's lush vegetation and animal life are the ones that the first Polynesians saw, when they discovered the islands many centuries ago; many are not even the ones that greeted Captain Cook a thousand years later. About a third of all species of native birds have disappeared since that first arrival of Europeans, but the original island-hopping colonists from Polynesia had already exterminated an even higher proportion of the bird species they found.

Of course, Hawaii is still beautiful enough to enchant every mainland tourist who sees it—which, perhaps, proves what we said early in this book: that these "Class 1" processes are only disturbing esthetically or morally.

Not all extinctions are wholly undesirable. (Can you think of anyone who mourns the passing of the disease organism that caused smallpox?) And there are some occasions when even honest and reasonable persons may wonder whether it is worthwhile to try to preserve some insignificant creature that finds itself standing in the way of progress. The famous "snail darter" incident of some years ago is a case in point: environmentalists, acting under the Endangered Species Act, fought through the courts to halt construction of an immense dam project because it would destroy the only known habitat of these tiny, undistinguished fish.

Of course, the real question in the snail-darter case wasn't so much whether that little fish was worth preserving as whether that particular dam had ever been worth building in the first place.

Before we leave the dismal subject of the destructive things we've done to the planet we live in, let's take a parting look at a place where almost everything has gone wrong at once . . . for it may be the shape of the future.

That place is the city of Venice, Italy, the celebrated Pearl of the Adriatic.

Venice is still a tourist Mecca, but after 1500 years of glory, the ravages of industrialization, overpopulation and tourism have nearly done it in in the last fifty. Pollution of the surface water? The Venetian lagoon is a poisonous soup of wastes from refineries, warm water from the cooling systems of power plants, domestic sewage and filth of every kind. (Venice has no sewer system. The toilets flush directly into the canals.) Over-pumping of the water in an underground aquifer? Venice has done it so well that the city has settled some five inches since World War II. In addition, the rising of the Adriatic Sea—due to the warmup? No one knows that for sure—has raised the sea level another three or four inches in the same time, which is why so many Venetians so often have wet feet. At the beginning of this century the Piazza San Marco flooded ten times in an average year; at the end of it, about forty. When there is a storm surge as well as a tide, the Piazza—the lowest point in Venice, and perhaps for tourists the central point for sightseeing—can be four feet deep and the Venetians cross it on board trestles. Or they row.

The eutrophication of Venice's famous lagoon is close to terminal. Rafts of algae cover such large spans of its water that it has to be harvested regularly, as much as five hundred tons a day. (Then the algae has to be trucked away and disposed of.) When the algae dies, its decomposition steals oxygen out of the water and so the fish die, too. The algae is fertile breeding ground for swarms of mosquitoes—so thick at times that, according to one authority, "airplanes can't fly at the airport and trains can't run, because the tracks are slick with dead bugs." The stink of all this is sometimes overpowering.

Yet even Venice is still beautiful—so much so that only half the people in Venice on any given day are Venetians; the other half are tourists from all over the world, drawn to its beauty and historic charm. It isn't likely that all of the rest of our world will be as fortunate.

All of these things we have done to our Earth . . . but we haven't stopped with our own planet. As we will see in the next chapter, we have even begun to pollute space.

12

The Pollution of Space

It's hard to believe that empty space itself can be polluted by human efforts. It's true, though. We have trashed the orbits our spacecraft travel through. So much so that, before every launch of a space shuttle, NASA's high-speed computers run for a full twenty-four hours on just one element of the flight: to choose a safe orbit, so that the shuttle won't destroy itself by colliding with some other man-made orbiting body.

Space is in fact fairly empty. But it is no longer empty *enough*. As of the last announced count (in early 1989) there were 7,119 man-made objects in Low Earth Orbit big enough to be tracked by surface radar. A handful of them are working satellites—communications, intelligence, weather. Some are satellites which have run out of energy or lost communication and are now "dead"—but remain in orbit because the laws of orbiting ballistics don't give them any other place to go. Some are pieces of scrap metal from broken-up satellites, launch vehicles or fuel tanks. A few are simply objects that the astronauts or cosmonauts have dropped—a wrench, a screwdriver, a Hasselblad camera. They range from the size of a baseball to the size of a school bus, and they are moving at high speeds—around four miles a second, fast enough so that even the smallest of them could seriously damage any object they happened to hit, even the shuttle.

Taken all together, what these abandoned space-borne objects amount to is a sort of orbital minefield, left there by the spacefaring nations of the Earth. They are space junk, and their number is still increasing. Two German scientists— Peter Eichler and Dietrich Rex of the Technical University of Braunschweig—have estimated that the chance of a "catastrophic" collision in space is now about 3.7% per year, and predict that if the present rate of increase of space debris continues, by the middle of the next century any such collision could set off a chain reaction.

Is there any way of cleaning up this garbage belt? There have been plenty of proposals for doing so, ranging from passive "vacuum cleaners" (a windmill-like satellite with huge plastic vanes, or a huge, miles-across ball of sponge plastic, either of which would simply soak up all the smaller objects in their path) to active robot spacecraft, remote-controlled, that would be flown ahead of the Shuttle through its orbit, as tanks shelter advancing industry. None of these is likely to be put into practice for three reasons. First, they are all terribly expensive. Second, none of them would work as well as simply keeping the trash out of orbit in the first place. And third, most of them might actually produce *more* small particles after a collision.

For those large objects are only the beginning. In addition to the big ones, there is an uncounted multitude of tinier objects in those same orbits, too small for the radar search at Goddard Flight Center to detect. There are somewhere around forty or fifty thousand of these smaller things from the size of a marble up—as well as a much larger number of tinier objects still. Perhaps there are a million of the least bits of trash: flecks of paint from old fuel tanks, fragments of metal tinier than a fingernail, odds and ends of litter of all kinds.

Even the tiniest of these bits of cosmic shrapnel are dangerous to anything they collide with. Small as they are, they are *fast*. Because of their velocities the kinetic energy—which is to say, the destructive power—of even the least of them in a head-on collision can do as much damage as a cannonball at sixty miles an hour. It is certainly enough, for instance, to kill an astronaut if he were unlucky enough to have his suit punctured by one.

Here, too, there is no end to the technological fixes that have been proposed. If the space station is ever built, for example, NASA has already made plans to surround it with a double wall of thin sheet aluminum—hopefully to absorb the kinetic energy before it hits anything important. Two engineers, Cyrus Butner and Charles Garrell, have patented a "Method and Apparatus for Orbit Debris Mitigation" which consists of a honeycomb of cone-shaped buffers lined with some energy-absorbing substance, and there are dozens of other ideas which have been put forth more tentatively.

Whether any of them would work is unclear. That they would add mass to every launch, and thus reduce the amount of payload that could be carried, is unquestionable. In any case, even if they worked there are many cases where they couldn't be used. The mirror of a space telescope or the receptors of many kinds of instruments simply could not function with any sort of shield between them and the things they are launched to observe.

The danger from these tiniest bits is no longer theoretical, either. It has already been proven to be enough to destroy a working satellite.

Consider the case of the Solar Maximum satellite.

Solar Max was launched in 1980 as a scientific instrument, charged with measuring the radiation from the sun. That is important research, because its purpose was to study the indispensable ultimate source of energy possessed by the human race.

Solar Max did its job very well for a while, but the good time lasted only for a few months. On September 23, 1980, the data from Solar Max stopped coming, and its ground controllers could not tell why.

For three and a half years after that date, Solar Max circled the Earth in silence.

Then, in April, 1984, the shuttle astronauts repaired it. They did it while both they and it were in orbit, and that was a truly remarkable feat. First they had to find the dead satellite and maneuver the shuttle to approach it as closely as possible. Then one of them had to put on a spacesuit, launch himself out into space to Solar Max, open the satellite up, remove the damaged parts and replace them with new ones. All of this had to be done while floating in empty space, more than a hundred miles above the Earth. It was a complex and exhausting job, and the astronauts did it beautifully.

They were successful. At once Solar Max came alive again. It took up its interrupted task; its reports began to come back down to the ground stations, and they kept on doing so until at last, five years later, Solar Max came to the end of its working life. It wasn't any kind of instrument failure that finally did it in. It was atmospheric friction. In the normal course

of the solar cycle the Sun's heat had warmed the Earth's atmosphere to the point where thermal expansion brought air
molecules up to Solar Max's level. The drag of the atmosphere slowed Solar Max down until it fell out of orbit. It
finally crashed into the Indian Ocean, near Sri Lanka, on
December 3, 1989.

What was it that had knocked Solar Max out of service for
three and a half years of its working life?

The answer to that came when NASA's scientists studied
the broken parts. The delicate instrument panels had been
riddled with tiny holes, 150 of them in a surface area about
the size of a card table. Solar Max had been struck with a
blast of cosmic buckshot. Some of those tiniest bits of orbiting junk, impossible either to see or to avoid, had collided
with it and killed its instrumentation.

So the risk of collision with orbiting trash is real. It has
not happened to just that one satellite, either. At least three
others are known to have been, or are suspected of having
been, damaged the same way; and, on at least one flight, so
was the space shuttle itself.

On the third day of its July, 1983, flight, the shuttle
Challenger—yes, the same one that was destroyed a few years
later when it exploded after launch—was hit with an object
too small to be seen, but big enough to pit the glass in the
pilot's windscreen with a crater the size of a pea. Again,
NASA's analysis showed the culprit: *Challenger* had been
struck by a tiny flake of white paint, no doubt chipped off
some old booster. Fortunately it had been only a glancing
blow, but even so the screen had to be replaced (at a cost of
$50,000) before *Challenger* could fly again.

So the number of objects in the trash belt in space

continues to grow, and most of them will remain in orbit for decades or even centuries, and there isn't any technological fix in prospect.

Worse still, if we should be unfortunate enough to see some current military projects grow to fruition—Star Wars, for instance—that orbiting minefield may fence us in so thoroughly as to prohibit future space projects from being accomplished at all.

Star Wars—officially termed the Strategic Defense Initiative, or SDI—was the bill of goods Edward Teller sold Ronald Reagan in 1983. It was supposed to put a "nuclear umbrella" over America; its backers invested a fortune in TV commercials, showing a sweet little girl sleeping in perfect security, with the Russians presumably gnashing their teeth in thwarted rage.

The opinion of most qualified scientists—at least, of those not employed by the project itself—is that there's no hope in the world that Star Wars can ever fulfill that promise. Even the project's own authorities have now cut their claims back, saying only that it can probably be used to protect some of our own missile launch pads against many of the ICBMs that would be launched against them. (The cities where sweet little girls sleep will have to take their chances.)

Whether even that claim is true, and whether that sort of limited defense is worth its staggering cost, is questionable. Some experts suggest that, in fact, Star Wars is more likely to provoke a nuclear attack than to prevent it . . . yet the project won't die. In spite of everything, in spite of the changed relationships between the USSR and the USA, in spite of all the evidence that Star Wars is a bad idea, it still continues to be funded with billions of dollars every year. 22% of the increase in federally funded research and development from 1983 to 1989 went to Star Wars-related projects, as did 11% of *all* American R&D, public and private. Although, before the war in the Persian Gulf, the future of many new high-tech weapons systems was in some doubt, with much talk of cancelling them for a "peace dividend," Star Wars had a charmed life. Then, of course, the swift and total American military victory in the Gulf revived all the high-tech warriors' hopes. The Patriot missile, the use of laser-guided "smart" bombs and many another new wonder weapon was hailed as the direct result of Star Wars research, and a justification for all its efforts.

None of those claims, of course, were true. Just how successful the new super-weapons were is hard to assess; all reports were censored by the military, and often colored for their own purposes. (For instance, although the bulk of naval missiles were launched from cruisers and destroyers, almost the only ones allowed to be reported were those launched from submarines and battleships—because the Navy wants them funded.) But to use a laser to guide a bomb is nothing like using a laser to blow up an enemy missile; and the Patriot (which was originally designed to attack enemy aircraft, not missiles, because it is totally useless against anything more sophisticated than the slow, clumsy Scud) was actually commissioned long before Star Wars was even a gleam in Ronald Reagan's eye.

All the same, the high-pressure public-relations of the military and the defense contractors has certainly obscured the issue. It now seems highly likely that at least some elements of Star Wars will be built and put into orbit within the next few years.

What will the effects of that bad idea be on the ecology of Low Earth Orbit?

We can get some idea by looking at past history—for instance, at the sorrowful story of the satellite called Solwind.

The astronomical research satellite designated P78-1—called "Solwind" for short—was launched from Vandenberg in February, 1979. The principal instrument Solwind carried was a coronagraph—which is to say, a telescope equipped with an "occulter" to block out the direct light of the Sun, so that the solar corona can be observed.

For six and a half years Solwind did the job it was designed for, churning out its pictures of the Sun's corona; it even did more than had been hoped, in fact, for example discovering a whole new class of sun-grazing comets. At times there were other science satellites observing the Sun. Some of them— the aforementioned Solar Max for one—were larger, newer and more sophisticated. But for the three and a half years while Solar Max was out of commission, Solwind was the only dedicated source of coronal data scientists had.

It was Star Wars that finally killed Solwind.

The Star Wars people have a history of conducting meaningless public-relations "tests" to make the taxpayer feel they're getting somewhere. Some of the tests are silly, if not even fraudulent ("destroying" a target with a laser, without mentioning that the target was surrounded with mirrors to concentrate the radiation; "targeting" one of the shuttle flights with a radar impulse—as though an enemy missile would be as obliging as the shuttle in announcing its orbital plans). However meaningless, all of them are hyped as major breakthroughs. When, in 1985, they needed another "breakthrough" to keep the appropriations flowing, they decided to destroy a satellite in orbit.

They picked old Solwind as the victim.

They made it a point, of course, to tell the world that the satellite they had chosen to zap was not only obsolete but no longer working. Of course, neither statement was true. Nevertheless they went ahead, pulverizing the Solwind satellite on September 13, 1985. The scientists who had depended on it for data protested vigorously, but of course the Star Wars people simply ignored them.

Solwind doesn't deliver any data any more. Still, it isn't entirely gone. Of those 7,119 trackable trash fragments in orbit, about a hundred are the remaining blown-up pieces of Solwind.

If, against all common sense, Star Wars is sooner or later even partially deployed, the trash belt in Low Earth Orbit will be multiplied many times over—even if no orbital war is ever actually fought. For instance, if the Star Wars X-ray laser (which requires nuclear blasts for power) were ever deployed, it would necessitate a hundred or more nuclear reactors in orbit.

That isn't likely to happen; even the Star Warriors seem to have given up on the X-ray laser. Currently their best bet for *some* kind of orbital defense is the "brilliant pebbles" scheme proposed by the Lawrence Livermore scientist, Lowell Wood. Each pebble would be a complicated and expensive satellite about ten feet long, weighing 350 pounds, guided to attack the first enemy missile it sees by its own on-board computer—a very *good* computer, comparable to a Cray-1, but less than a thousandth its size, filled with (as Wood says) "so much prior knowledge and detailed battle strategy and battle tactics" that "it can perform its purely defensive mission with no external supervision or coaching."

That is to say, the computer will decide for itself what to attack, and when. No human being will control it. (No such wonderful computer exists at present, either, but the Star Warriors are always hopeful.) And there will be thousands of these smart pebbles in orbit.

As we have seen from the studies of the two German scientists above (which were presented at the 1989 International Astronautical Federation meeting in Torremolinos) just adding all those additional bodies in orbit may make collisions almost inevitable, after which one big smashup could produce so much space junk that it could set off a chain reaction. It is not clear how many brilliant pebbles and associated hardware, even if never fired at an enemy in combat, could bring the number up to that point.

But some things are clear, and one of them is what happens if Star Wars actually is deployed, is committed to combat and actually works.

What happens then? How many tens or hundreds of thousands of trackably-large hunks of junk (never mind the millions of tiny ones) would be left in orbit from the use or destruction of all those communications and control satellites, missile casings, even nuclear charges; of popup lasers, smart rocks or brilliant pebbles and all the rest of the wonderland of high-tech, high-mass gadgets the Star Warriors wish to set spinning around the Earth over our heads?

How many centuries would it then be before a space program is possible again, without unacceptable risks of collision with some of those hurtling hunks of scrap?

Our space exploration has not merely trashed the Low Earth Orbits, it has done a job on the surface and the atmosphere of the planet as well. The two solid rockets for each Shuttle launch burn seven hundred tons of ammonium perchlorate. The chlorine this contains makes acid rain, for which reason the Shuttle is only permitted to launch when the winds are blowing offshore, to keep the stuff from falling on Florida's cities and farmlands.

The chlorine in it may also contribute to ozone-layer destruction, and the fuel, which is a hundred and sixty tons of aluminum, definitely does. It produces aluminum oxide. That chemical isn't naturally found in the upper air, but when we put it there it produces particles which are just the right size to form "seeds" on which the ice crystals that facilitate ozone destruction can form.

Elsewhere in the world, the Soviets are right now trying to clean up several million acres of land in the district called Dzhezkazgan, just east of their rocket launch pads at the space base of Tyuratam.

Rockets have to be launched toward the east, to take advantage of the boost they get from the rotation of the Earth. The Soviets didn't happen to own a usable eastward-facing coast like Cape Canaveral, so they launched from the middle of the continent. That meant that the 890 launch-stage boosters launched from Tyuratam fell to the ground in the Dzhezkazgan region, where they still remain. They contain pumps and other parts which hold toxic fluids, so much of them that the soil there is now too contaminated to allow cattle to graze on it.

Of course, our American space program didn't have quite the same problem. Our own boosters and toxic chemicals all fell into the Atlantic Ocean. The difference, of course, is that ours are not visible, and so can't be cleaned up.

The space program is, by and large, one of the most wonderful and promising endeavors ever undertaken by the human race. It would be folly to abandon it. But it is even greater folly to fail to exercise caution in how we carry it out . . . if only because if we don't, we may reach a point when we can't have a space program at all.

THE
TECHNOCURES

13

What Can We Do About It All?

That completes the bill of particulars. If we haven't listed every last thing we human beings have done to make the world worse for most living things, ourselves included, at least we've hit the highlights. We don't have to cover every last detail.

Probably this is a good time to do a little more reality-testing. Start by asking that same hard question again:

Do we really *need* to change our lives around because some scientists say things *look* bad? After all, it's still possible that they made a mistake in their arithmetic, or misunderstood what their observations meant, and so things may not really get as terrible as they expect. So—do we act? Or do we wait and see?

That's a tough call to make, but as it happens it's the same tough call many of us are faced with when we go to the doctor and, after he's made all the tests and studied all our symptoms, he tells us that we're sicker than we thought, and we need an operation.

That's not the kind of news we like to hear, and we would like to think the doctor is wrong. The first thing we are likely to do is to start looking for a second opinion—or a fifth, or as many as it takes to find somebody who will say we're really pretty healthy after all.

In the real world we live in, there are plenty of environmental doctors—some of them scientists, some of them political leaders—who will be glad to tell us that we don't have to worry about a thing.

It's even possible that in some cases they're right. But it's rather more likely that they're not. So prudence would suggest that we do what we generally do when the doctor tells us something we don't want to hear, and that is to grit our teeth and start taking the cure.

So . . . what's the cure for the world's environmental ills?

There isn't a single cure, of course, because there isn't a single ill. But we've already seen that some of the ills seem to be related. Acid rain, the greenhouse effect and air and water pollution, for instance, all seem to be linked with the burning of fuels for energy. So one cure might be to go on an energy diet.

That sounds unattractive enough to begin with, because few of us are really anxious to give up, or even cut down on, the comforts and pleasures that all that energy provides for us—things like our cars, our air conditioners, our electrical appliances. (And those large sections of the human race who don't now possess those things—the poor populations of the Third World—are certainly going to want them just as much as we do. It will not be easy to try to deny them what we already have.) All those wonderful toys are totally useless without the energy it takes to run them, and most of that energy comes from the burning of fossil fuels.

Can we reduce the damage from fossil-fuel burning and still keep at least most of our comforts?

Fortunately there's a happy answer to that. Yes. We can.

In fact, we have a great many options open to us in our quest for ways to make this world habitable for our grand-children. Some of the measures we can take will be trouble-some. Some, however, will be quite easy, because all we have to do in those cases is to quit doing some of the things we do that we gain no benefit from anyway.

Let's face it, we Americans are power pigs. We aren't the *most* profligate wasters of energy per capita in the world (our neighbors to the north may be startled to learn that Canada occupies the Number One spot in that area), but we're in the top three or four.

There's no reason for it, either. We don't *need* all that en-ergy. One proof of that statement lies in the fact that such other countries as Germany and Japan are just as well off in every measurable way as we are (as we can easily deduce from the way they beat the socks off us in the export marketplace). Yet they use only about half as much energy per capita as we do to accomplish their economic miracles.

The reason they can do that is simple. They're smarter than we are.

At least, they're more prudent in their energy consump-tion. They use exactly the same kinds of energy we do, for exactly the same purposes, but they use it more *efficiently*.

There's nothing to keep us from learning to copy their ways. In fact, it would save us a great deal of money. Our own American experts, like Amory Lovins, have been telling us that for years. According to Lovins, the savings we could achieve by using more efficient lights, cars, heating systems and machines add up to our greatest unexploited energy resource, far larger than anything we could hope to pump out of the ground by despoiling every square inch of Alaska and all of our coastal waters. It is just about equal to the total amount which would be generated by all the new power plants the American public utilities are planning to construct over the next twenty years. It is enough to make a real difference.

That's not the end of the list of things we can do. There are even a few important actions which, as a matter of fact, can be taken quite readily (though certainly not without some serious resistance from parties whose profits would be affected) because what is at fault turns out, on examination, to be little more than bad bookkeeping.

We'll see how that is so a little later on. First let's look at the facts in the case—the available "technological fixes" which we have learned generally to despise, but which in this case do offer hope. And, since energy is at the root of all these measures, let's start by looking at how our power plants work, and in what ways we can make them work better.

14

Power

When we turn a switch in our homes we expect the lights to come on. They almost always do.

The electricity that lights American homes and runs American business comes from an immense interconnected network of public-utility power plants. The generating plants are scattered all over the United States and Canada, but their output is joined in a large, complex series of grids.

Some of the power plants are hydroelectric, getting their basic energy from falling water. A tiny fraction of the others generate electricity from other renewable resources, such as windmills or solar energy. Every other power plant—far the greatest majority of them, worldwide—is basically a heat engine of some kind. A few burn gas and run it through turbines directly. All the rest are steam engines.

These plants burn some kind of fuel to make heat. The heat boils water into steam. The steam expands through turbines, where its thermal energy (the energy of heat) is turned into kinetic energy—the energy of motion. The turbines turn generators, and electromagnetic energy—electricity—comes out.

Even the nuclear power plants are steam engines, for the only use they make of their atomic reactions is to heat water to a boil. However, what the great majority of power plants burn is fossil fuels: coal, oil or natural gas. In the process they produce the acid-forming gases that make acid rain and the carbon dioxide to trap the heat of the Sun and bring about the global warming. The nuclear plants don't do those things, but they produce their own undesirable by-products: spent nuclear fuel, too radioactive to be kept around but without any good place for disposal, plus the odd Chernobyl or Three Mile Island accident.

Let's look at how this process works, step by step, in tabular form:

FUEL
is burned to make
HEAT
which, by producing steam, is turned into
KINETIC ENERGY
which in the generators becomes
ELECTROMAGNETIC ENERGY
in the form of electricity.

Those are the four basic steps in the process of generating the power to run your toaster—plus a lot of highly sophisticated engineering to get the maximum electricity out of the fuel.

So in order to see how we can generate the electricity we need (or at least that we want) without the damaging by-products, let's look at these steps one at a time, and try to see if we can eliminate some of their problems.

Step 1: Fuel

There isn't any *scientific* reason why the fires in our power plants have to burn fossil fuel like coal, oil or natural gas. There are other things to burn.

Some of these other fuels are in use right now in the handful of power plants which burn "biomass"—which is to say things like wood, crop wastes, organic matter of many kinds, even municipal trash. These aren't as convenient to handle as fossil fuels. The biomass fuels often have a high content of moisture, so some of the heat they produce when they burn is dissipated into drying them out. They don't generally contain as much sulfur as fossil fuels, which is a plus. On the other hand, they sometimes produce more chemical pollution (in particular that's a risk with burning municipal trash) and all of them do produce just as much carbon dioxide.

However, as far as the global warming is concerned, that carbon dioxide doesn't really count. There's a significant difference between burning biomass and burning fossil fuel. As the name suggests, fossil fuels are *buried*. Coal, oil and natural gas don't exist on the surface of the Earth. They can't, because if they did they would have burned spontaneously, or slowly oxidized, long ago. As long as they remain unmined their carbon content is locked away from the air.

Biomass, on the other hand, exists on the surface of the Earth, in contact with the atmosphere. Even if it isn't burned, it slowly decays and forms carbon dioxide anyway. That slow oxidation may take years or even longer—but fossil fuels *never* do that until we mine and burn them.

Another advantage of burning biomass is that it is a good way to get rid of some kinds of waste. Sawdust and chips from lumber mills are a common fuel. So is agricultural waste, for instance the "bagasse," or spent canes, from sugar mills; the island nation of Jamaica now routinely burns bagasse for electricity, and if that were done worldwide it could generate some 50,000 megawatts, the equivalent of fifteen or twenty coal-burning plants. Almost any crop residue can be used as fuel, though some of them, leafy materials in particular, must be compacted and pelleted first.

A splendid source of biomass to burn in power plants is municipal garbage—though one now fraught with dangers, due to our bad habit of allowing into our garbage collections plastics and other chemicals that produce toxins when burned.

Still, a little later in its life cycle that same garbage can produce another fuel for us to use. When garbage decays in landfills it emits methane gas, which is an excellent fuel but a great nuisance in every other way. No one likes having that constant garbage reek seeping out, so a few enlightened municipalities have dug collection pipes into their garbage dumps and burned off the methane to generate power. Methane comes out of decaying animal dung, too, and when farmers collect the excrement and pile it in concrete sheds called "digesters" it too can then be burned. Such enterprises are too small for public utility use, but they do work; the People's Republic of China has four million of those excrement digesters in operation now, using the methane gas they produce for cooking in the farmers' households.

And, of course, the more methane we burn the less there is left to seep into the atmosphere, where—as we have seen—it plays a significant part in trapping solar heat and bringing about the greenhouse warmup. (Isn't it nice that there is this happy feature of some conservation plans that they accomplish several desirable things at once?)

In that connection, there is another good idea waiting to be implemented. We've been talking about accidentally acquired biomass from waste so far, but there's no reason we can't grow the biomass we need on purpose.

Years ago, a scientist at Emory University in Atlanta came up with an interesting plan. His name was Pong, and he looked at several of the problems then facing the nation. Carbon dioxide was not yet seen to be one of them, but there were plenty of others: the need of cities to get rid of sewage; the erosion and degradation of much cropland; the unemployed farm workers forced off their land; and, most of all, the energy crunch—it was just about the time of the OPEC oil embargo and there was an urgent need for new fuel sources. The Pong plan was simple: Pipe the cities' raw sewage onto the played-out farmland, thus restoring it for crops. The kind of crops it could then grow would be limited—no one would really want to eat tomatoes or lettuce grown with raw sewage—but quick-growing trees and shrubs would flourish. They could be tended as needed by the unemployed farmhands and, finally, they could be burned to replace imported oil.

Like many other innovative ideas of the time, the Pong Plan disappeared without a trace when oil got cheap again, but similar plans are being revived now. There are plenty of woody shrubs that could be farmed for fuel purposes; some scientists refer to them as "BTU bushes." Once again, they could confer many benefits. There is a lot of farmland that shouldn't be cropped any more, because it is being worn out, but is still farmed for the sake of agricultural subsidies.

It would do well for growing fuelwood, and the farm-support savings would take a little of the pressure off the budget deficit. The woody shrubs need not be burned just as they come off the farm; they can be converted through gasifiers into burnable gas, or through fermentation into alcohol— another excellent fuel, and one which (as we will see) could even be used to run our cars. And one scientist, Peter de Groot, echoes Pong by pointing out that if "BTU bushes" were planted on a large enough area to make a real difference—it would have to be quite a large area, in fact about the size of India—they could stop a lot of soil erosion and thus restore some badly damaged waterways. At the same time, they would add so much new vegetation which would be working to take the excess carbon dioxide out of the atmosphere that they would, at least to a small extent and temporarily, slow the global warmup down.

Of course, nothing is perfect in this world, and the burning of biomass for fuel presents two problems. The first is physical: burning biomass in large quantities produces some methyl chloride, which is one of the ozone destroyers. Fortunately it doesn't produce *much* methyl chloride, so on balance it's still worth doing. The other problem is moral, and applies only to farmed biomass: Many scientists are against the idea of farming for fuel when the world's growing population will need more and more food from its farms.

Probably what that means is that only *some* land, not very good for producing food crops, should be used for growing biomass fuel. Even that would help; and anyway there is still all that organic waste that can be burned instead.

* * *

The one significant remaining fuel alternative for "steam engine" power plants is nuclear.

As everyone knows, nuclear power has risks. Any power plant can have an accident, but nuclear accidents are worse. If a boiler in a coal-fired power plant explodes it may wreck the plant; but if a nuclear plant explodes, as Reactor No. 4 at Chernobyl did in 1986 and as several others nearly have, the damage, and even loss of life, can reach to places hundreds or thousands of miles away, even on other continents.

Chernobyl is the most famous nuclear power plant accident, but it isn't the only one. Other plants have had explosions and meltdowns; the difference is that Chernobyl's blast released large quantities of radioactive poisons into the world's air, where most of the others managed to be contained. In addition, there have been a *large* number of very near misses. In the United States, there was Browns Ferry (where a maintenance man was sent to plug a leak in the insulation; he had trouble finding it and so lit a candle to see where the flame was drawn—which set fire to the building and nearly destroyed all the safety features), a plant just north of New York City (where a discharged employee decided on arson as a form of revenge—fortunately the fire he started was controlled before it reached the nuclear reactors themselves), the March, 1990, near-disaster at the Vogtle nuclear power plant in Waynesboro, Georgia, when a truck backed into a utility pole and cut off power to the control systems of one reactor (fortunately the reactor was shut down for servicing at the time), a number of accidents at military reactors making nuclear warhead materials and, of course, the famous Three Mile Island. In the former East Germany there was an even closer call in 1975; a fire in their Greifswald reactor destroyed all the interior power cables. One single pump, which happened to be connected to an outside power source, was all that kept the reactor from a catastrophic meltdown and a Chernobyl-style disaster.

The nuclear power industry knows all about these accidents, of course (though they do not go out of their way to share that knowledge with the rest of us). They're doing their best to prevent them. The nuclear engineers have been working hard to try to make the existing nuclear power plants safer, and, farther in the future, there are much less dangerous designs for nuclear power plants that are still on the drawing boards.

The best of these possible future power plants start with wholly new designs. Most of the world's existing nuclear plants, by an accident of history, follow the basic design originally created for America's nuclear submarines.

The nuclear submarine plants were the first serious, large-scale effort to get useful power out of nuclear reactions. They worked quite well. Since then, because public utilities are a lot more comfortable scaling up and improving known technologies than pioneering new ones, most commercial electrical generators are basically those same old submarine plants, though of course greatly modified and improved.

There's a limit to how much you can improve an old design, though. New-model power plants, cut loose from their military antecedents, could be a lot less accident-prone.

One of the most promising is the "modular high-temperature gas reactor" or MHTGR. It uses the gas helium rather than water as a coolant; the hot helium then heats water into steam for the turbines. This means that there is less difference in temperature between the fuel and the coolant, which helps achieve full passive safety—which, when you translate it, means that if anything were to go wrong (the people at General Atomics claim), the reactor could go on harmlessly working at full power for days, while repairs were made.

Other systems rely on gravity-fed water reservoirs. What keeps the nuclear reaction under control is water cooling, in a system not unlike the radiator in your car. In present designs, that water has to be pumped, and pumps can fail; then you face the real chance of explosion or meltdown. In the proposed new system, if the pumps fail gravity will keep the coolant running into the core for some time.

Still other systems propose using a different sort of radioactive nuclear fuel. The kind of fuel now used in most reactors is uranium carbide, which when exposed to water produces a chemical reaction. It gives off the explosive gas acetylene. (That's why many nuclear-power plants have a burnoff fire constantly going, like the gas flares at an oil refinery, as the accumulating acetylene is piped away and burned.) With a different fuel formulation that dangerous constant flow of acetylene gas would not be produced in the first place, and so one more risk of explosion would be eliminated.

That's the good news. The bad news is that, no matter what happens, very few of those new designs will be in service very soon.

It takes something like ten years to complete the engineering research before groundbreaking for a major power plant, and then an additional eight or ten years of construction time—even if the government or someone else provides the necessary research funding (several hundred million dollars a year for the next ten years) to complete the studies that might make these new fail-safe designs practicable.

Nor does that solve some major environmental problems: even if you are sure your new nuclear plant won't blow up, what do you do with its radioactive wastes?

One possible (though dubious) hope for avoiding that problem lies in that other great new idea, the widely publicized future hope of fusion power.

Fusion power is a whole other kind of atomic energy. There are two known ways of releasing power from the nucleus of the atom. One is fission: splitting up very heavy atoms, like uranium, and that is what every nuclear power plant in the world does now. The other is by *combining* (or *fusing*) light atoms, like hydrogen; that's what makes the hydrogen bomb work.

Fusion power would be wonderful in many ways if we had it, not least because the fuel it burns can come from that vast reservoir, the water of the world's oceans. For that reason, a great many scientists have been working very hard for several decades in the attempt to find a safe and economical way of putting that fusion reaction to work in a commercial power plant. They've discovered a lot of interesting science. They've thought up a number of different ways of attacking the problem. They still hope—at least, most of them do—that all the problems can be solved, and hydrogen fusion will provide cheaper and cleaner energy.

But not now. Don't don't hold your breath. It won't happen soon. Fusion-power enthusiasts have been promising to deliver it within twenty years since the 1950s. Now, almost forty years later, it is still twenty years in the future.

Actually, fusion power may not happen ever—a good many qualified scientists think there is no good way, even in theory, to make fusion power economically worthwhile—and it certainly won't happen in time to prevent the global warmup that is facing us now.

Those are the major fuel alternatives before us. Some of them could make a large dent in our problems if we began using them on a large scale.

But we can do even better than that. Remember, fuel is only the first step in the power-generating process. The only reason we burn it is for the sake of creating the next step, namely heat.

So what about the alternatives *to* fuel? Why do we need to burn anything at all?

Step 2: Heat

Since we only burn fuel to make the heat to boil the water to make the steam to run the turbines, let's look for other sources of heat. Some of them don't involve burning of any kind, chemical or nuclear.

There's an unlimited supply of one such source, for instance, right under our feet. It's called geothermal energy.

Geothermal energy is the heat that comes from the interior of the Earth itself. That central part of our planet is *hot*. In fact, the core of the Earth is mostly composed of melted rock and iron.

There's no way for us to drill down to the molten core, even if we wanted to, but good old Gaia has provided us with an acceptable substitute. There are plenty of places where the hot rock comes very close to the surface—sometimes in the form of volcanoes, more frequently showing up as hot springs.

Some of these gifts of geothermal heat energy are already being exploited for commercial power. On sites in Iceland, California, New Zealand, Italy and elsewhere in the world it is practical to drill into underground pools of hot water, and this has been done. The utilities there pump the hot water out of these high-temperature aquifers to the surface, where it expands into steam to turn generators.

Other places have no such hot aquifers, but they do have hot subterranean beds of rock. In those places it is possible to make the hot water, which you do by pumping surface water down into the rock until it is hot enough, then pumping it back up for power generation. Sites of this kind have been identified in literally thousands of places around the world, including even such unlikely locations as Antarctica (Mount Erebus would be a good prospect, if the time ever comes when human beings want to live on that frigid continent) and the British Isles. (Though the British, after some preliminary drilling, decided there were easier ways to get renewable-resource power and abandoned their geothermal project.)

Even geothermal power is not without some penalties. The installations do some damage to the landscape (though not as much as oil wells), and along with the hot water some radon may come up from underground. Radon is a radioactive gas, and exposure to it can cause cancer. (For that reason, as we'll discuss later on, some householders need to test their inside air and arrange for venting if they find radon present.) But the health hazard would not ordinarily be very significant in a geothermal plant. Any radon that came up with the hot water would be vented into the open air, where it is quickly dispersed. And because radon has a relatively short life, it disappears entirely before very long.

Another available source of "fuelless" heat is the waste heat from industry.

A steel mill, for example, burns a lot of fuel to melt its metal, and most of that heat is then allowed to waste itself into the air; why not use it to boil some steam up to drive a generator and make electricity? This rarely used resource of waste industrial heat has been available for a long time; what has kept it from being exploited on a large scale is partly the engineering problems involved, which are not really very complicated, and even more the extent to which public utilities have traditionally resisted any threat to their monopolies. But now government intervention and public pressure have compelled some changes, and so this "co-generation" is beginning to be exploited, with significant prospects.

Finally there is the free and very large supply of available heat from the Sun.

Solar power is another form of nuclear-fusion power. The best thing about it is that that particular nuclear-power re-actor is located a comfortable 93,000,000 miles away from us. The energy from the violent nuclear reactions at the core of our star comes to us in the form of sunlight, immense floods of it, everywhere on the daylight side of Earth.

How much energy is that? Put it this way. Less than a tenth of one per cent of that solar energy gets caught and uti-lized by plants, but it is that tenth of one per cent that sup-ports all life on Earth. In total, the energy which the Earth receives from sunlight *every fifteen minutes* is equal to the aggregate of the electricity produced by all the world's power plants in an entire year.

The trouble with sunlight is that it is *low-density* energy. You know that for yourself, because you know that you can leave a tank of water out in the sunlight forever without generating any power. The heat will slowly evaporate it, all right, but what it won't do is boil into useful steam.

But the answer to that is only a question of engineering: all you have to do is somehow concentrate the heat, as generations of Boy Scouts have done with their magnifying glasses to start campfires.

For electricity generation the engineering is more complicated. A magnifying glass won't do the trick, but other devices will. In the late 1970s the Sandia National Laboratories in New Mexico built one such power plant. They called it the "Solar Power Tower," and it consisted of a large number of cheap mirrors, on rotating bases so they turned as needed to follow the Sun across the sky, all aimed at a boiler on the top of the tower. The concentrated reflected sunlight struck the boiler; the boiler turned water into steam. (A later generation was planned to omit the water-into-steam cycle entirely, using the great heat available to produce electricity directly through the phenomenon called magnetohydrodynamics or MHD.) The Sandia tower was only a test and produced no electricity, but a prototype of a commercial power plant of that design later went into service in California and did contribute 80 megawatts of electricity to the California grid. (It too was subsequently abandoned, because it was not efficient enough to be competitive, but the principle is established.)

Farther in the future is the proposal to put something

like the Solar Power Tower in orbit, converting the sunlight
to microwave energy which would be beamed down to
receivers—called "rectennae"—on Earth.

Getting our household electricity from space in that way
would solve a great many problems, and scientifically it's a
quite reasonable idea. Practically, not quite so reasonable.
The capital costs are daunting, even if we had a vast and reliable fleet of thousands of shuttle-like spacecraft to haul all
that construction material into orbit, which we don't . . . so
it is not likely to happen for quite a while.

Perhaps the most imaginative attempt to generate electrical power from the Sun's heat is the Ocean Thermal plan
tested out in Hawaii.

The main problem with using solar energy is building
something to collect the sunlight for us. The bigger the collector, the more energy we collect to use. In the Ocean Thermal plan, the collection area is not only large enough for
almost any purpose, but is provided to us without cost by
nature: it is the surface of the Pacific Ocean itself.

After all, the sun continues to warm the upper layers of
the ocean every day. As we pointed out above, the water
won't boil, but its warmth can be used in another way. In the
Ocean Thermal design, this warm surface water is used to
heat a sealed system of some fluid which has been chosen to
vaporize at a temperature of less than a hundred degrees
Fahrenheit—the temperature of the surface water. As the
liquid turns into gas, it is passed through low-pressure turbines, producing electricity. At the same time, cold water
from deep in the ocean is pumped to the surface; when the
gas comes out of the turbines it is cooled with the cold water
from the depths and becomes liquid again . . . after which it

passes back through the system to be warmed into gas again and recycles indefinitely.

There are other possible designs for extracting usable heat from sunlight. They all have great advantages: none of them use any fuel at all, they go on working as long as the Sun shines, they do not pollute the air or leave undesirable wastes of ashes or nuclear residues. They all also share one disadvantage. They are capital-intensive. They require a considerable investment to get them started and as long as interest rates are high and fossil fuel is cheap the utilities prefer to build more coal plants.

As we have intimated before, there are other ways of keeping the books on power plants, and with a different system of accounting that cost advantage may disappear.

But, even without revising our accounting systems, we can find other technologies for producing electricity that not only don't use fuel but don't even use the heat we burn the fuel for. After all, the only reason we want the heat is to turn it into kinetic energy—that is, to make something *move*—in order to turn the turbines. So why not just look for sources of free kinetic energy?

Step 3: Kinetic Energy

It was free kinetic energy that began the Industrial Revolution a century and a half ago. It came in the form of flowing rivers and waterfalls, and it was put to use by inserting a bladed wheel into the moving flow of fluid.

When the moving fluid is water—a cataract or a stream— it is called waterpower, and very soon after large-scale commercial generation of electricity became possible waterpower began to be used to turn the generators. The product is "hydroelectric" power. In the United States such massive enterprises as Hoover Dam and the Tennessee Valley Authority were based on cheap hydroelectric power half a century ago, and they revitalized whole regions of the country with millions of inhabitants.

To generate hydroelectric power, all you need is a quantity of water at one level, and a nearby lower level for it to flow into. As the water seeks its own level it descends from the first point to the second. Along the way you can use it to turn your generator turbines.

Some sites are better for hydroelectric purposes than others. If no good natural site exists, it is possible to create an artificial one by building a dam. Technically speaking, a natural site like Niagara Falls is practically perfect for the purpose, because its collection area is four of the five Great Lakes; no dam needs to be built, and no new reservoirs created to drown out good land. (Niagara's one handicap is that it is so beautiful. The amount of Niagara water devoted to generating electricity is rationed at a low level so that the scenic splendor of the Falls will go on attracting tourists.)

The worst thing about generating electricity from hydropower is that so much of it is already being done. The natural sites in the United States are all being exploited already, and building new dams involves drowning a lot of useful, or merely beautiful, land areas. The most promising hydroelectric sites in the United States have been used already, so there aren't many good sites left for dam-building anywhere in the country (except, possibly, in Alaska). Indeed, some of the dams that have already been built have probably done more harm than good. Some of the rest of the world is in better shape in that respect. There is considerable untapped hydro-electric potential at sites in the Himalayas, Iceland, South America and Africa, though most are not likely to be developed very soon; they are remote from the heavily populated areas that need the electricity most.

But hydroelectric power is not our only source of "free" kinetic energy—free, that is to say, except for the expense of building the devices that can capture it.

Whenever something is in motion, energy can be extracted from it, and one continual stirring of motion in the world is in the air itself: wind power.

Windmills are a familiar technology—think of all the windmills that were employed in grinding corn in Holland or pumping well-water to the cattle herds in the American south-west. In modern times wind power has even been used to generate electricity for the power grids in New England—until about 1950, when it became cheaper to buy and burn fossil fuel than to keep up the maintenance on the last big wind-mill.

Wind power is only useful in places where there's plenty of wind, but that situation is common on the tops of many hills or along most coastlines, as well as on the great American plains. At this moment, in the hills near Livermore, California, there are nearly seven thousand wind turbines at work, and about as many more again in other parts of California. Collectively these wind turbines generate 1350 mega-watts of electric power—a third more than the nuclear reactor which exploded in Chernobyl in 1986—and there is room in America for many more. About 20% of the present U.S. demand for electricity could be met by properly sited windmills, using innovative designs now being tested. The price is pretty nearly right, too. It is estimated at about 7 to 9 cents per kilowatt-hour, easily competitive with nuclear-power plants (though, according to present accounting procedures, not with the most efficient fossil-fuel ones).

The ocean itself is in constant motion. Anyone who has ever wandered along a seashore has seen the thousands of tons of water that rise and fall with the waves. The British have probably done more research on wave power than most. One proposed system deploys a number of floating "ducks" near a shore, their bobbing action as the waves pass under them converted into energy for making electricity.

Besides the waves, there is one other constant movement in the sea: The tides. They, too, can contribute. In certain areas local geography produces a funneling action so that tides at the end of a bay may rise and fall fifteen feet and more. Most areas are not as generous with their tide heights, but that can be dealt with by amplifying them. One amplification scheme calls for excavating lagoons on the landward side of such shores so that water will flow in as the tide rises, and out again as it ebbs. In all of those cases you have sizable volumes of water in motion, which can be used to turn a turbine on both the rise and the fall.

But even this kind of kinetic energy is still only one more way-station along the road to the electricity we want. Perhaps we can jump right to the goal direct.

Step 4: Electromagnetic Energy

You see, we start out in a flood of electromagnetic energy every day. All light, including sunlight, is simply one special form of electromagnetism; electricity is just another form of the same basic force. If we can convert the one into the other we can skip all those intervening steps of fuel, heat and kinetic energy. Then we can just pull all the kilowatts we need out of the light of the Sun.

The way to do that is through the devices called photo-voltaics, or PVs.

PVs use the photochemical effect (the same thing that used to open a supermarket door for you with an "electric eye") to transform sunlight directly into electricity.

Theoretically that process is elegantly simple; in terms of engineering, full of problems. Some of the chemicals that do the work of transformation are sensitive to one particular wavelength of light, some to another. So in order to get the maximum efficiency the photovoltaic elements have to be built in layers, each layer picking up energy that the one above it passed by. It has not been easy to find substances that are efficient enough, and transparent enough, and sturdy enough (so they don't have to be replaced as weather and time deteriorate them) for the job.

The answer to that is research. The general theory is well understood. All that's needed is the kind of research and de-velopment to make it more effective.

Indeed there has been a great deal of research on photo-voltaics already, but there's less of it going on now than there was a few years ago. This isn't because the scientists are discouraged. It's because during the Reagan years the annual budget for PV research was cut from $150 million to $35 million—largely because of the influence of the fossil-fuel powers—and the cuts have not been restored.

Even so, photovoltaics are already a commercial success in some applications. PV is the cheapest and best way to supply relatively small quantities of electricity to places which are remote from the utility grids—for example, to provide power for vaccine refrigerators and water pumps in Third World countries. European manufacturers of equipment for aid programs do use photovoltaics for that purpose. American manufacturers do not, because they're not allowed to by law. U.S. government aid agencies insist, for whatever mad bureaucratic reasons, that all such devices continue to be powered by conventional fossil-fuel burning generators.

When, and whether, photovoltaics can take over any substantial share of public-utility electricity generation we cannot say. They never will, unless the funds for research are restored. The government does spend a lot of money on energy research—some $40 billion a year in the 1980s. But it spends its money in the wrong places. Nine-tenths of that government research money during the Reagan-Bush years has gone to fossil-fuel and nuclear power research—both of which spend immense fortunes on political lobbying and influence-buying—and only 4% to research on producing energy from renewable resources.

We have not described every system for generating electricity. For example, it is possible to convert Step One, Fuel, direct into Step Four, Electricity, without ever going through Steps Two and Three. That can be done by means of such devices as the phosphoric acid fuel cell, or PAFC.

This device runs on hydrogen. The PAFC can be made in almost any size and would be particularly valuable for generating relatively small amounts of power in remote locations—say, enough to run guest houses or a small hotel in a national park or on an offshore island—since it wouldn't need expensive transmission lines to connect it to a full-scale power plant. And if a renewable source of hydrogen could be obtained—say, from photolysis of water by sunlight—it too would be a renewable resource.

But all of these renewable-resource systems for generating useful electricity have five features in common, three good, the other two bad.

The first good feature, of course, is that *they do not destroy our environment*. That is the most important of all, and it nullifies any possible disadvantages.

But there are plenty of other really good aspects to the use of renewable resources. In addition to being in fact really renewable (we will never run out of wind, waves or subterranean heat), they have zero fuel cost. Since they don't require any imports, there is no fear of OPEC embargo-type blackmail; or of helping to finance the war chests of people we may find sooner or later fighting against our own soldiers, like Saddam Hussein; or of overstraining the balance of payments: there's no need to buy fuel from anybody at all. Once the renewable-resource plants are operating, the electricity they produce is, apart from maintenance and distribution costs, essentially free.

The other good feature is that, although they are capital-intensive, they are also relatively quick and easy to construct.

This is a very large practical advantage, often overlooked. It typically takes time on the order of a decade from go-ahead to actual electricity production for even the simplest fossil-fuel plant. As we have seen, nukes take even longer. While almost any renewable-resource plant can be designed, built and put into operation in a period ranging from a few months to no more than two years. This means extremely important dollar savings, because less money has to be borrowed for long, and expensive, periods of time.

There remain the two negative features of most renewable resources.

One of those is that renewable resources can't always be turned on or off at the will of the power plant operator. If the wind doesn't blow, the windmills won't turn; after sundown photovoltaics are idle; when the ocean is calm, the waves don't roll in.

That is a real problem, but not one unique to renewable-resource systems (even conventional utility systems have times when some units are down by accident or for maintenance). What is more important, several good solutions already exist.

The first solution is to have a mix of energy sources, not just one. Even when it is dark the winds will likely still be blowing; if the winds fail, the tide will still be coming in and out. There will almost always be a substantial number of generators turning over somewhere in the system, and since all power plants are connected through the long-range grids it matters little which ones are contributing the needed electricity to the net.

The second solution is power storage.

Some hydroelectric plants already store power by tapping the head of water to run their generators when electricity is in peak demand, then reversing them (using electricity from other sources in the grid) to pump some of the water back up into the reservoirs for future use when the demand is low.

There are other technologies in development. Huge storage batteries, like the battery in your car but with hundreds of thousands of times as much capacity (as much as 8.5 megawatts), have already been built in places like West Berlin,

Germany, and Monmouth County, New Jersey. They are expensive and, at present, not very efficient, but they work. Energy can also be stored kinetically (huge flywheels set rotating at off-peak times, then used to turn generators when demand is heavy), or as compressed air (the system called CAES, or Compressed Air Energy Storage, which pumps air into an underground cavern, to be released as needed to turn gas turbines) or in many other ways. And, of course, if worst came to worst and there just wasn't any spare energy stored anywhere, one could always start up some mothballed oil-burning plant.

That still leaves us with one discouraging feature of renewable-resource power systems: Money.

In spite of all their advantages, most of them can't at present compete economically with nuclear or fossil-fuel power plants.

The difference in cost per delivered kilowatt-hour is often quite narrow, and is decreasing with time. New designs are always coming along (though not as rapidly as they would if the government backed the renewable-resource research with as much money as it spends on other energy problems). The machines become more efficient. But it still costs more on your utility bill to get a kilowatt-hour of electricity from a solar-thermal plant or a tidal generator than from a conventional fossil-fuel plant. (Though often not as much as from a nuke, and we have many of those chugging away all the time.)

That situation is bound to improve. According to estimates from the Union of Concerned Scientists, wind-turbine electricity, which now costs about 7 to 9¢ per kwh, will become available in the reasonably near future at no more than 4 to 6¢ per kwh with improved designs. Solar-thermal power, now about 12¢ per kwh, should drop to 8¢. Photovoltaic, the most expensive of all, now costs from 25 to 35¢ per kwh, but that price should fall to 6 to 12¢.

But that's all in the future. What's more, that future will take an unacceptably long time to arrive unless the United States government does come to its senses and put the necessary money into the necessary research.

However, there's a side to this question of costs that is ignored in almost all calculations. Most of the extra expense of renewable-resource electricity is not real. It is a fiscal illusion, if not an actual fraud practiced on all of us.

As we will see in the next chapter, the "impractically high" price the "experts" talk about really represents nothing more than sloppy bookkeeping.

15

How the Bookkeepers
Can Save the World

We are going to start with two facts about fossil fuel prices that would surprise most Americans.

You see, none of us really know just how much we pay for fuel—especially fossil fuel (oil, gas, coal) and the things we buy that come from them, like electricity—because bad bookkeeping hides the truth from us. The first fact is that sometimes the price we pay is far less than it needs to be. The second fact is that often the price is a lot more than we think it is when we hand over the money to pay the bill.

It might seem as though these two facts might more or less cancel each other out. They don't. What the effect of the way our fuel prices mislead us actually results in is our being cheated in both directions.

To see how this works, and to see how proper bookkeeping standards could save us all a lot of grief, let's start with the check you send each month to your local utility company.

Since utility companies are in business to make money, when they send you your electric bill they calculate it accordingly. The amount you pay includes your share of all the utility's out-of-pocket production costs (fuel, payroll, overhead, capital amortization, maintenance, etc.). To that sum they add a profit to give to the utility company's stockholders.

That bill they send you is very carefully calculated by the utility's accountants, and audited by government authorities. But all the same it's for the wrong amount.

You see, you don't pay all the costs incurred in the production and distribution of your electricity. What's more, your utility company doesn't pay them, either.

The price we pay for our power is a fake. There are some very high costs involved in the production of electricity which no utility company ever shows on its books at all. They are what are called the "external costs."

In the language of accountancy, an external cost is any expense incurred in the course of an operation which the proprietors of the operation don't have to pay for themselves. The price of generating electric power includes a lot of these external costs, but they do not show up in your bill—or in the electric company's.

For example:

It certainly costs money to steam-clean an office building which has been dirtied by soot from the burning of fossil-fuel in a power plant. It is also true that the people who operate the electric plants which generate that soot never pay that cleaning bill. Nothing could be farther from their minds. The people who own the building are the ones who pay this particular external cost of electricity, as they always have. Then, as they always have, they pass these costs on to their tenants in the form of higher rent—and then those tenants pass them on to their customers (like you and all of us) as higher prices for whatever they sell.

The bill gets paid. Such bills *always* get paid. But not by the right people.

Nor do the plant operators pay the costs associated with the carbon-dioxide trapping of solar heat that threatens to upset the global climate, or the costs that arise from the acids and toxic chemicals that erode the noses off marble statues, kill fish and trees, rot rubber automobile tires, damage crops, make people sick and, yes, even *kill* people. All of those expenses are levied on the world at large.

Similar costs arise from many other processes than electricity generating. Home heating furnaces contribute their share of these external, but very real, costs. So does the manufacturing industry in general, with its toxic emissions and effluents. Automobiles contribute a *huge* share of these heavy costs, both in their manufacture and above all when we get behind the wheel and drive them, as we'll see in the next chapter . . . and few or none of these off-the-books expenses are paid by the people who cause them.

All in all, the people of the United States pay out literally billions of dollars each year to meet these otherwise unmet "external costs" of fossil-fuel burning of all kinds. Nor is the rest of the world spared the obligation of paying such bills. In fact—as we have seen—a good deal of the rest of the world, particularly in such highly polluted places as Eastern Europe, is stuck with far higher "external cost" bills than even we are.

So, if we want to calculate the true cost of, for example, fossil-fuel generated electric power, we have to internalize these external costs. In the case of electricity generation, that means the bookkeepers must add the external costs onto the conventional costs of running the power plant.

The easiest way to do that is by imposing a "carbon tax" on the electricity.

Right there we sense a sudden cooling of enthusiasm. More *taxes*? you ask indignantly. Who the devil wants to have to pay more *taxes*?

Nobody does. But the fact is that governments are going to levy more taxes no matter how we taxpayers feel about it, because that's where they get the money to keep them going.

So don't ask whether you want more taxes. Ask instead whether you want the new taxes that you're inevitably going to get hit with anyway to do some *good*.

We think there's only one answer to that. So let's look at all the ways in which such a tax may be more bearable than most—and a lot more useful, besides.

A "carbon tax" is just what it says it is: it is a tax on the amount of carbon that will be burned in fossil fuel. It doesn't matter whether the fuel is coal, oil or natural gas. They all contain carbon. They all contain hydrogen, too, and those two elements are the constituents of the fuel that we actually burn to create energy; but burning hydrogen doesn't pollute in the same way so the hydrogen part of the fuel doesn't matter.

So the first step is for chemists to tell us what percentage of a barrel of oil, say, is carbon—that's very easy—and then for the legislators to put a tax on it. How much should the tax be? That's a harder question, but then questions like that are what bookkeepers are for. For the moment let's say that it amounts to some arbitrary amount like a dollar a barrel.

That means that instead of costing say $20 a barrel, all the oil used in this country will then cost $21. And then, when that cost is passed on, it is reflected in the price of the electricity we use (the part of it, anyway, that is generated by burning that kind of fossil fuel), the gasoline we burn in our car and so on.

That increases the cost of all those things to all of us. But look at what we get for our money.

A carbon tax accomplishes three very good things. First, by increasing the cost of fossil-fuel generated power, it will cause people to economize on their use: thus the less fossil fuel that is burned the less carbon dioxide and other pollutants are generated to go into the atmosphere. Second, by reducing or eliminating the price difference between fossil-fuel and renewable-resource power, the carbon tax encourages the use of non-polluting renewable resources. Finally it provides money that can be used for many good purposes.

All those things are valuable. None is enough to meet our needs by itself, though.

A mere price increase would not put an end to that kind of pollution unless the tax were made huge. We saw that at the time of the Persian Gulf crisis. There was a 50% price escalation for gasoline almost overnight, but the sales of gasoline dropped only marginally; people complained loudly, but they paid the extra cost and kept right on driving.

How huge is "huge"? Some estimates have been made. A British group calculated that a 3% carbon tax, gradually increased by an additional 3% each year thereafter, would stabilize our carbon dioxide emissions by the year 2005. Similarly, the American Office of the Budget estimates that a $28 per ton tax would cut our carbon emissions by 20% by the year 2000 (and raise about $40 billion a year in revenues).

But other estimates are a lot heftier. Some studies suggest

that the price of fossil fuels would have to be *quintupled* to reduce consumption enough for us to meet the 20% reduction in carbon dioxide emissions recommended at the 1988 international Toronto conference on the subject.

That would be a considerable shock. If any such 400% added tax were imposed immediately the consequences would be dramatically nasty—especially for poor people, who would face serious problems in, for example, heating their homes to even a barely livable level in winter. Some sort of subsidy would have to be provided to keep them alive—and, of course, every such subsidy would encourage others, and all of them combined might seriously vitiate the purposes of the tax.

But, whatever the details, some kind of substantial and increasing carbon tax would do much good. It would make it economically worthwhile for industries and utilities to switch to alternate sources, as well as to spend the money necessary to increase energy efficiency. It would do the same for individuals like ourselves, especially as our aging homes, cars and appliances became due for replacement. It would encourage prompter measures as well: a homeowner will find it worthwhile to invest several thousand dollars in adding insulation to his house if he knows he will save several times that amount in fuel costs over the next decade.

There is a troubling effect of most new taxes, including a carbon tax, and that is on employment. Government statisticians estimate that 78,000 jobs disappear with every dollar-a-barrel increase in oil prices. But that sort of effect is temporary; the same estimates say that by the fifth year of a major effort to limit the burning of fossil fuels employment would begin to gain rapidly as more efficient use of energy produced cost savings and stimulated the economy. The savings wouldn't be limited to energy costs, either. For governments (and thus for taxpayers like all of us) much money could be saved by taking the long-range effects of a carbon tax into account in planning, for instance, future highways: with less car use, less expensive roadbuilding would be necessary.

In some parts of the world, the first "carbon tax" may soon become a reality, for in September, 1990, the German environment minister, Klaus Topfer, proposed to tax all smokestack industries at the rate of 10 Deutschmarks (about $5) per ton of carbon dioxide emitted.

Of course, carbon dioxide is not the only problem. We also face the need to do something about acid rain and, although the carbon tax would help in that respect simply by cutting down on the amount of fuel burned, we may need also to consider a separate levy on acid-forming chemical emissions, whether they come from factories, utilities or automobile exhausts.

So, whether we impose a carbon tax to help slow the greenhouse warming or not, we do very badly need a tax to cover some of those other externals.

These costs are very hard to measure. No one can at present put a trustworthy dollar value on the cash loss to all of us caused by pollution damage to vegetation and public health. Still, we do have some tentative estimates.

One preliminary study suggests that the appropriate tax on fossil-fuel electricity—the amount that would be equal to the costs of the external costs to the world caused simply by the poisoning of the air—would be in the neighborhood of three to four cents per kilowatt-hour.

That turns out to be an extremely interesting figure, which opens up a whole new world of environmental possibilities.

Such a tax would not merely punish those who pollute. Added to the current costs of electricity, that amount would then produce a situation in which it made good economic sense for the polluters to change their procedures in ways which would stop that sort of polluting entirely.

If the external costs of generating fossil-fuel electricity were internalized by a tax in that range, it would instantly wipe out much of the economic disadvantage of renewable-resource systems. There would be no economic reason for public utilities to postpone the switchover to non-polluting energy sources. Since there is no real reason *but* the economic one, they might even bite the bullet and do it, for then solar power and wind-power generation in particular would begin to look like real bargains . . . as, indeed, they are.

In the same way, a tax equal to external costs on gasoline would have a comparable effect.

If anything, the identifiable external costs incurred every time a gallon of gasoline is burned in a car are even higher than with power generation. Those costs don't stop with the pollution that comes out of the exhaust pipe. If the cost-accounting took *all* the external costs of automobiles into consideration, it should also include the heavy financial burden of providing roads and parking spaces for all those cars, not to mention the loss of fertile land paved over to provide them, or the cost in human lives from traffic accidents.

There, too, if any part of the external costs were internalized into an added cost per gallon at the pump, they would have a powerful long-range effect in causing individuals to switch from cars to public transportation, in encouraging the use of more fuel-efficient cars and ultimately in designing non-fossil-fuel cars.

Does that sound simple? Of course it does. It has to sound that way, because it is simple.

It's not only simple, it's also quite fair. An added "internalization" tax doesn't compel anyone to do anything in particular, nor does it interfere with anyone's civil liberties. We all would retain the right to do pretty much what we liked. Those who want to highball across the country at ninety miles an hour—wasting gasoline and pumping out fumes— can go right on doing it, at least as long as the state police don't catch them speeding. Those who want to floodlight their homes like a Christmas tree will be allowed to burn all the electricity they like.

The only thing that would change is that they'd have to pay for their excesses, and what could be fairer than that?

Now we come to the second fact that most Americans don't know: the fact that we pay a lot more *in cash*—not counting all these external environmental costs we've been discussing— than we think we do. For example, if you ask any American what a gallon of gas costs him he will most likely quote the figure he last saw advertised at his local filling station . . . never imagining that what he actually pays in his own hard-earned money is many times that.

The extra price is paid in taxes, and it's paid in several ways.

One way is simply in government handouts of our tax money to the oil industry. The American government heavily subsidizes all fossil fuels, oil perhaps most of all. These government handouts take the form of tax credits, research grants and outright subsidies, and they are huge. The direct out-of-Treasury dollar payments to fossil fuel producers alone amount to some $26 billion a year; what the additional un-justified tax breaks they get come to in total is obscured—not accidentally—by the way the oil interests have managed to get the legislation written, but must be many times that.

That applies to all the oil we consume, domestic or foreign. The imported oil, however, costs us heavily in a quite different way. We pay for it in our military budgets.

This isn't just a question of the costs of the actual war we waged in the Persian Gulf in the beginning of 1991. According to an analysis by Harold M. Hubbard (published in *Scientific American*, April, 1991) even in 1989, well before Saddam Hussein showed any signs of wanting to invade Kuwait and thus to bring about a shooting war, the United States spent between $15 and $54 billion on patrolling the Gulf with its Navy and other military measures in that area. (If the amount is fuzzy, it is because it is possible to argue about just what budget costs should be charged to which particular ledger item . . . but even at the minimum figure the amount is large.)

These measures had only one purpose. They were explicitly and admittedly for the purpose of protecting the sources of imported oil. It happens that very little *American* imports came from either Iraq or Kuwait, so in essence the price we were protecting was quite often the one that was charged to our competitors in the arena of world trade . . . but we were the ones who paid the bill.

What it comes down to is that every time you take a $10 bill out of your pocket to pay the filling station for the gasoline you put in your car, you are in reality spending an additional $20, $50 or even $100 per fillup that is concealed from you.

And that is in addition to all the environmental costs we've already described—some in dollars (the "externals" we've listed), some in loss of environmental benefits (remember the *Exxon Valdez*, etc.), some in human life (not just the damage to health caused by pollution, but the actual deaths each year in, for instance, coal-mining).

So what *do* you pay for your fossil fuels? And how much longer do you think we can all go on paying it?

We can't solve all our problems with environmental taxes, but if we keep our books properly, and tax where taxing is needed, we can certainly make a dent in the problems.

We did say a bit ago that such taxes are both simple and fair. That's true but, of course, "simple" and "fair" don't mean the same thing as "easy."

Apart from the instinctive revulsion any good American feels when he hears the word "tax," there are serious problems with putting any such taxing plan into practice. There will be nothing easy about it.

Taking *any* effective action to preserve the environment requires persuading people to give up a present convenience—in this case, cheap energy—for a future good, namely the preservation of the world for our grandchildren.

Experience teaches us that that will be a very hard sell. At best, if we are going to impose environmental taxes it means that somehow we are going to have to induce legislators to pass laws that many of their most influential constituents will loathe—and will punish the legislators for, by withholding contributions to their campaigns, and by supporting their opponents in elections.

Here we get into the thorny area of politics. We can't avoid it. We'll talk in detail about how hard these political efforts will be later on in this book—and how, if we want to, we can accomplish them anyway.

But before we do any of that we need to look in more detail at some other great running sores in our environment, starting with our cars.

16

Getting There:
Cars, Trains, and Planes

The biggest single source of pollution in the world today is transportation. The human race burns tremendous amounts of fossil fuel to get itself and its goods from one place to another, and the outstanding favorite way of getting around is in the family car.

We Americans in particular love our cars.

We know it's sinful. We have been told over and over that every time we start our cars up we are making pollution. We've been informed that the ten million barrels of oil we burn in them each day come out of the exhaust pipes as a vast poison-gas smog of unburned fuel droplets, carbon monoxide, organic chemicals which sunlight turns into ozone, acids and toxic chemicals; and that cars and trucks produce 25% of American carbon dioxide, 75% of its poison-gas brother, carbon *mon*oxide, 45% of airborne hydrocarbons and above half of all the toxic chemicals in the air.

That doesn't stop us. We do it anyway. That's love.

We all share that love, regardless of age. Every teenager's first ambition is to get a car of his own. Growing up does not terminate the devotion. For adults, the car is a toy, a pet and a status symbol. City dwellers will get up at five in the morning to beat the parking regulations by moving, from one side of the street to the other, the car that they may only actually drive on an occasional weekend. The Mercedes or BMW owner considers himself to be just a little bit better than the person who drives a Ford or Toyota, and the sports-car person lives in a snob heaven of his own. We pay very heavily for this indulgence—even in actual out-of-pocket dollars, as well as the environmental costs. One dollar out of every five the average consumer spends goes into his car. Altogether, cars and the businesses associated with them account for a startling 10% of the American gross national product.

As Jay Leno puts it, "You can take away Americans' TV, you can even take away their guns, but you'll *never* get their cars away from them." For many people, the car *defines* the American Way of Life. Nor is this passion limited to the United States.

All the same, our cars are killing us.

These car deaths don't happen just by violent collisions, as in the tens of thousands of Americans who are killed and the hundreds of thousands who are maimed in auto accidents in America every year. The cars are also poisoning the environment.

So we don't really have a choice. Something has to give. However difficult it may be to make major changes in our relationship to the car, some real changes are *necessary*.

Of course, some changes have already begun. Now there are laws to (somewhat) reduce the amount of poison a car pumps into the air we breathe; catalytic converters are mandatory now for that purpose. The converters work fairly well—for the first 50,000 miles or so. After that they lose efficiency rapidly. Anyway they do nothing at all about the greenhouse carbon dioxide (almost a pound of it for every mile you drive) which the cars emit.

By all means the best way to cut down the quantity of poison gas from our cars is to cut down on the amount of gasoline we burn in them.

Three good strategies are available for us to accomplish that. The most radical of them, and in the long run probably the one we will wind up with, is to replace today's gasoline with some other—and non-polluting—fuel.

We'll see what we can do about that a little later on, but right now the kinds of cars that can use the alternative fuels well aren't on the market, so we'll have to concentrate on the other two strategies.

One of those is to use the cars less—to car-pool, consolidate errands, take public transportation, ride a bike or even walk. The other strategy is to improve the gas mileage we get when we do drive them.

Improving gas mileage is entirely practical. We know that because when, in the 1970s, the OPEC oil embargo made gasoline conservation a necessity, the American Congress did take some steps in that direction and they worked. One of the laws it passed then—a markedly unpopular one—did in fact produce instant and large fuel savings.

That law was the federal 55 mile per hour speed limit imposed all across America. The law was widely violated, true, but it *worked*.

It may at first seem that slowing down the traffic on our highways would not be very important in saving gasoline and reducing pollution. What makes it so is a different kind of law, the natural laws of physics: *Speed costs gasoline*.

The reason for that is that about a third of the gasoline your engine burns is spent merely on the job of shoving the air in front of your car out of its way—that is why "streamlining" has some advantages for high-speed travel. The faster you go, the higher that proportion of energy used to move the air ahead of you becomes, since it is a fact of natural law that air resistance rises as the *cube* of velocity. A car going twice as fast doesn't encounter twice as much air resistance. Instead, it has to force its way through the cube of two, or *eight*, times as much resistance.

What that comes down to in terms of typical highway speeds is that your car engine requires slightly more than twice as much energy to overcome the air drag at 70 miles an hour as it does at 55.

That doesn't mean that the car burns twice as much fuel. There are other drags on the car—such as internal friction in the engine itself and in the transmission, as well as rolling friction between wheels and road surface—and air resistance is only one factor which affects mileage. What it does mean is that if you drive a hundred miles, and restrain your foot on the gas pedal to the lower speed, you will get there a little later but you will have burned much less gas on the trip.

(As a fringe benefit, you also may save your life. Accidents also rapidly get more damaging as speeds increase. In fact, they kill. A New Mexico study showed that when the federal speed limit was relaxed in 1987 the fatality rate almost doubled. 550 *more* people died as a result of car crashes that year, in New Mexico alone, than would have died if the limit had been retained.)

Unfortunately, the savings from the federal speed limit did not produce the hoped-for real savings in the total amount of gasoline burned. We found a way of beating the system. We just bought more cars, and then we drove more miles on all of them. Americans add an extra four million cars and trucks on the road each year, and the rest of the world is right in there with us. The number of cars in the world has been doubling every ten years since 1950, from a mere fifty million then to a staggering 400 million-plus by 1990. And, of course, the world's oil consumption has had the same identical eight-fold increase—not coincidentally—reaching more than 200 billion gallons in 1990.

(And, of course, the more miles we drive the more support services we need—trucks and snowplows to make the roads passable in winter, never-ending use of construction machinery all year around—and all of those machines produce their own pollution as well.)

There are other ways to improve gas mileage—technological fixes—and Congress tried to compel some of those to be adopted at the same time.

Legislation authorized the Environmental Protection Agency to set a timetable for car manufacturers, requiring that the average mileage per gallon be improved, step by step, over a period of years. For the first few years that worked fairly well. The manufacturers complained with great passion, but by and large they complied with the directives anyway, so that the cars of the 1980s were on average somewhat more fuel efficient than the ones that had gone before.

Then oil got cheaper again, at least temporarily.

We've reversed the trend to fuel economy. Under pressure from drivers Congress nibbled away at the 55-mph limit; under even greater pressure from car manufacturers and

owners, the EPA manufacturing standards that had been
set in the 70s were relaxed in the 80s. The Reagan administration didn't think that saving fuel was important. Congress, weak-kneed as always, didn't want to go up against a
popular president, so it obligingly slowed its timetables for
the car makers. As a result the average fuel efficiency of the
1989 new-car fleet dropped from 27.2 to 26.8 miles per gallon. (And the fad for trucks, station wagons and four-wheel
drive vehicles, which average only 21 miles per gallon,
made it worse.) The more fuel-efficient cars certainly saved
on gasoline, but you couldn't hang quite as many energy-
using appliances on them. And, above all, they weren't quite
as *big*.

There is a limit to what Congress can do. In the long run,
it can't accomplish anything that is contrary to the national
will—which is to say, the aggregate of all the decisions of all
of us. If we want to have more fuel-efficient cars on the roads,
we'll have to be willing to buy them.

The car manufacturers say we won't. Their biggest argument against reform is that they complain they can't afford
to make energy-efficient cars because the public insists on
buying the big fuel-guzzling dinosaurs.

What the car manufacturers don't talk about is the *reason*
for much of that, namely the hundreds of millions of advertising dollars they spend on persuading consumers to want
them. That's the whole purpose of advertising, of course. All
those television and radio commercials, all those space ads
in newspapers and magazines and all the mail and phone so-
licitations, billboards and other instruments of selling have
only one function, and that is to make us want something
we didn't want until they made us want it.

* * *

There's a useful hint there for those of us who want to save the environment. It would be helpful if we could school ourselves to *resist* the advertising blandishments of those who want to sell us things we don't much need. If we did that we might even be able to forgo some (not all) of the gadgets we hang on our cars.

The gasoline we burn does more than just move the car along the road. It also provides the energy for lights, radio, air conditioner and all the myriad other accessories the typical car salesman of the 1990s wants us to buy. Each one of them takes energy to run—some only tiny amounts, like the lighted dials on the instrument panel; some a lot, like the air conditioner.

There are ways we can *reduce* the wasted energy, some of them quite simple. (Even just choosing the right color for the next car you buy can make a difference. As with your house, the lighter the color of the paint job, the less heat it conducts, and the less energy you have to spend on your air conditioner.)

Still, all that accessory energy has to come from somewhere, and the place it comes from is our gas tanks.

The accessories also take energy to build. The energy needed to build all those air conditioners and tape decks and motors for moving the driver's seat around produces its own pollution. Roughly speaking, the energy cost of building a car about equals the energy required to drive it for a period of a year or more, and contributes about as much in damage to the environment. The more accessories, the more manufacturing energy is spent.

From this general rule it follows, as the night the day, that you can reduce pollution simply by keeping your cars longer. (There's another advantage to driving a little slower. It helps in prolonging a car's life; you won't wear the things out quite as rapidly.) A good way to make your car last longer is to look for cars which are sturdily built in the first place, and for which repairs are made as easy as possible so the cars can be maintained.

We've already mentioned size as something to look for when you do buy a car. By and large, smaller cars are more fuel-efficient than large ones. That's a matter of simple physical law, too: it takes more energy to push two tons of metal along a highway than one.

To be sure, smaller, stripped-down cars are less fun to drive than big ones with every possible option hung onto them. For most Americans, the size of the car they drive increases in step with their increasing age and affluence; they buy as much iron as they can scrape up the installments to pay for.

Not every driver would admit that his preference for larger cars is a matter of pleasure or status. Some would say it is a matter of safety, since big cars are said to be more survivable than compacts in a collision, for instance.

That's true, as far as it goes. It is a fact that if you run a Cadillac head-on into another car, you are more likely to walk away from the accident than if you do the same thing in a Volkswagen Beetle: your crash is cushioned by a good deal more collapsing metal in front of you in the larger car. But that is only part of the story. The main reason a small car is dangerous in a collision is that the car it collides with is likely to be a massive big one—if not a fourteen-wheel semi truck-trailer. If there were fewer juggernauts on the roads, all compacts would automatically become safer.

Anyway, if we carried that kind of reasoning to its logical conclusion we'd all prefer to drive M-1 Main Battle Tanks.

So we can save a good deal of fuel, and thus help reduce the damage to the environment, by following those two strategies: by exercising self-restraint in the size and complexity of the cars we buy, and by limiting ourselves in the velocity and frequency with which we use them.

But self-denial isn't our only hope. Technology can help us a good deal in the long run.

For one thing, if we want to cut down on the pollution caused by burning gasoline, we might start looking for something else to run our cars on. There isn't any law in heaven that says we have to rely on fossil fuels to make our cars go. There isn't even any law that says our cars have to have a fuel tank at all.

You remember from our discussion of electric-power generation that the reason we burn fossil fuels in power plants is to convert the thermal energy in the fuel into the energy of motion that drives the generators. It's the same in our cars. This means that the same sort of shortcuts may be available to us here, too. If we could find a way to *generate* the kinetic energy the cars need to go in some other place, and then *store* it in our cars, then we can drive off without ever burning anything in the car itself.

Such "fuel-less" systems do exist. Auto manufacturers have already produced designs, and even prototype models, for cars which operate with no internal energy-generation at all.

In one design the car is equipped with a massive flywheel; the wheel is spun up to speed overnight, using electricity from the utility lines. Then, the next morning when the car is needed, a clutching device like the automatic-shift transmission in a conventional car allows the driver to tap that rotating-wheel kinetic energy to drive the car. In another plan, the energy is stored in the form of compressed air in a tank; again, the tank is pumped up overnight with public-utility power, and when the air is released it drives a turbine that moves the car.

Concepts of that sort don't require any true "engines" at all. Still, both those concepts have serious problems—at least at present they don't store enough energy for long-range driving, and they necessarily add a good deal of mass to the car. Stored kinetic-energy drives could well fill some niches in the overall vehicle mix, for example as short-haul urban delivery vehicles, but a better idea for general-purpose automobiles might be the electric car.

There is even one variety of electric car that doesn't need any fueling at all: it runs by sunlight, turned into electricity by photovoltaic cells on its roof. In an Australian road race in 1989 a flock of such cars were demonstrated; their speed was not great and the loads they could carry relatively small— but they do work. (Photovoltaic vehicles even now are practical in a few very specialized applications. Like golf carts, which don't have to go very far or very fast and have plenty of idle time to recharge their batteries.)

More common is the battery-driven electric car. These do already exist; in fact, they have for nearly a century. In the early days of the automobile they competed successfully with the gasoline-driven ones. They might economically compete still, if gasoline hadn't become so cheap. In some applications they still do: the United Kingdom right now has 35,000 electric vehicles on its roads, many of them delivering milk.

The present generation of electric vehicles has some problems. Because the storage batteries are heavy and limited in capacity, the electric cars don't go as far or as fast as gasoline-driven cars. Considerable improvements in that area are possible. Since 1980 a zinc and chlorine electrolyte battery has been available. In a light car, that battery would allow 200 miles of normal driving before recharge—the equivalent of half a tankful of gasoline. Then there is the new General Motors electric car called the Impact, which has a slightly smaller range—124 miles—but a top speed of 100 miles an hour, substantially more than most drivers ever need—or can use without going to jail. The Impact is not now in production. It won't be until General Motors is convinced the market warrants it—which is to say, when the company has any reason to believe that people like ourselves would buy it—but it too works.

In environmental terms, the advantage of all these plans is that the energy they use is produced externally—in stationary central power plants. Even if the energy came from burning fossil fuels in the first place there would still be less total pollution than from gasoline-burning, but we can do better than that. If those power plants used the kind of renewable resources we discussed earlier the cars would be essentially totally pollution-free.

Let's look, then, at the way to burn renewable-resource fuels in the engines of our cars themselves, with the same advantages for cars that the BTU bush has for power generation.

Automobile fuel, too, can be *grown*, rather than extracted from the earth.

Oilseed fuels right now are sometimes used to substitute for petroleum diesel fuel, pressed from plants such as sunflower, rape, hemp, soy, coconut and oil palm. They are not perfect for the purpose at the present time. They have a bad habit of producing residues which clog the engine, but there are technological fixes now under development which solve that problem (by pretreating the oilseed fuel with heat and catalysts which help minimize that damage). Some of these oil-producing plants are extraordinarily rich in fuel resources. The freshwater algae, *Botryococcus braunii*, is 85% oil (it is that oil content that makes it buoyant enough to float in water), and its oil can be refined to produce gasoline.

The physical facts are favorable to this sort of farm-produced fuel. Unfortunately, the economic facts aren't: farm-grown oils are not competitive with petroleum in price . . . yet.

Neither, at the moment, is hydrogen.

That's a pity, because hydrogen burns with as much energy as gasoline and produces essentially no pollution at all. When hydrogen burns with air at a controlled temperature (so that the nitrogen in the air isn't burned into nitrogen oxides at the same time), what comes out of the tailpipe is basically nothing but steam. The exhaust from a hydrogen car might make the heavy-traffic areas of our cities a bit more humid in the summer, but it would have no other discernable effect.

There are two problems that need to be solved before we can begin burning hydrogen to drive our cars.

The first is where we are to get the hydrogen from. If it is made, as most manufactured hydrogen is, from chemical reactions starting with natural gas, then the process of producing the hydrogen also produces some undesirable pollution. (We don't have to make hydrogen that way, though. If we did the things necessary to produce electrical energy without serious pollution, as discussed earlier, it would be possible to manufacture unlimited amounts of hydrogen, without pollution, through the simple electrolysis of water.)

The other problem is more serious. It is very difficult to store enough hydrogen in a car to run it very far.

Hydrogen happens to be the lightest of all gases. In order to carry enough of it to give a car a decent driving range the hydrogen has to be compressed drastically—which means strong, sturdy, and thus very heavy, fuel tanks. There are other possible ways of storing the hydrogen—in the pores of blocks of metal, for instance—but there are still unsolved technical problems in storing and releasing it when needed, and the heavy chunk of metal it is stored in imposes a considerable weight penalty there as well.

But there remains one possible non-fossil fuel which has few or none of the disadvantages of the other candidates, and indeed is now in considerable use around the world. That fuel is alcohol.

There are two great advantages to be gained by running our cars on alcohol. To date, the one that has been most decisively important in encouraging its use as fuel is the economic one. Alcohol does not have to be imported.

Not every country has its own oil wells, but there is no country in the world that can't manufacture its own alcohol, out of its own resources, from materials at hand. Thus there are no worries about incurring foreign debt or being dependent on foreign sources that may be cut off without warning.

The environmental advantage is in the long run more important still.

Alcohol-fueled cars produce far less in the way of acids, ozone and photochemical smog than gasoline cars.

This does not mean their exhaust is wholly innocuous. Nitrogen oxides may occur in the burning of any kind of fuel, alcohol included, since it is the nitrogen in the air itself that gets trapped in the combustion process. That is a problem which has available solutions, however, since (as we said above) the nitrogen oxide emissions can be controlled by careful regulation of the combustion temperature. Nevertheless, the burning of alcohol has a special polluting by-product of its own: it produces small amounts of formaldehyde— a serious poison, in fact a main constituent of embalming fluids.

Petroleum interests have had a lot to say about that. However, it is a paper tiger; the formaldehyde problem is really a lot less serious than it seems. It's true that the direct emission of formaldehyde from alcohol burning is more than from burning gasoline—but, paradoxically, the formaldehyde that then can be detected *in the air*, where it would do any harm, is actually less. The reason for that is that the bulk of the formaldehyde produced by burning gasoline—80% of it— doesn't come directly out of the exhaust pipe, and thus cannot be detected by monitoring emissions. It is produced subsequently by the action of sunlight on other exhaust chemicals. This doesn't happen with alcohol. All in all, alcohol is incomparably the cleaner of the fuels.

This is not a theoretical conclusion; it has been established by some pretty large-scale efforts in using alcohol to drive cars.

The United States is one country with such experience, for since the mid-80s an increasing proportion of motor fuel is no longer pure gasoline but "gasohol"—gasoline with some fraction like 15% of alcohol added. It works just fine. (In fact, American racing drivers *prefer* alcohol—every car in the Indianapolis 500 is alcohol-fueled.) Some American pleasure-car drivers try to avoid using gasohol because they're afraid it might make starting the engine harder in cold weather, but that's another paper tiger. Government tests showed no detectable starting problem with the 15% alcohol mixture, even in a Chicago winter.

Other countries have gone much farther than the United States, notably Brazil.

Brazil started its "Pro-Alcohol Program" in 1975, for the same reasons as impelled every other nation at that time to reduce oil consumption: it didn't want to be wholly dependent on OPEC. Brazil's choice of alcohol fuels was ethanol, or "grain alcohol" (though it was derived not from grain but from the fermentation of a large part of Brazil's sugar crop). Methanol, or "wood alcohol" (though this is not a very accurate name, either, since most methanol is now produced from petroleum or other non-wood sources), can also be used for automotive fuel, but Brazilian environmentalists don't like it; in fact, when the Brazilian government imported some methanol to eke out its ethanol supplies in 1989 the environmental forces forced the state fuel authority to reject it. (Not all the reasons were entirely environmental. Part of the popular objection to the use of methanol, rather than ethanol, lay in the fact that methanol is poisonous when drunk—

which put at risk every Brazilian who followed the fairly common practice of siphoning fuel out of one car's gas tank to put it in another's.)

Brazil's alcohol-fuel program moved very rapidly. In 1979 the first pure-alcohol car was manufactured in Brazil; by 1985, 96% of all new cars were alcohol-fueled; by the end of the 1980s, more than half of Brazil's autos and trucks used alcohol for their fuel.

However, problems began to come up, and by the end of the 1980s things began to go downhill for the Pro-Alcohol Program.

The economic advantages had dwindled. The price of imported oil had fallen as drastically in Brazil as anywhere else in the world, and besides Brazil had just discovered some substantial offshore oil reserves of its own. Worst of all, the Brazilian sugar farmers were in revolt. Although sugar was in a worldwide glut, the world sugar price was still high enough so that the farmers could get more for their crop by selling to sugar refineries than to the alcohol plants. That threatened fuel scarcities for Brazil, and the outlook for the 1990s was that all those alcohol-driven cars were going to be competing for an inadequate supply of the alcohol they needed to make them run.

Alcohol, it was found, had some other faults, too. The worst was damage to carburetors. Some Brazilian car owners complained they had to replace their entire car engines every two or three years because of the corrosive effects of the alcohol fuel.

That isn't surprising, nor does it threaten the future of alcohol cars. The problem in both cases is simply that the engines that are damaged are the wrong engines. These engines

THE TECHNOCURES

were designed to burn gasoline and only modified as much as necessary to accept alcohol as an alternative fuel. That's not good enough. To make the best use of alcohol as an automotive fuel the engine and in fact the whole car should be rethought and redesigned from scratch.

That process has already been begun. In 1989 two scientists from the EPA's Emission Control Technology Division in Ann Arbor, Michigan, published plans for a methanol car. Their names were Charles Gray and Jeffrey Alson.

Methanol burns at a cooler temperature than gasoline, so the Gray-Alson methanol car would take advantage of that fact: it wouldn't need to have any radiator or cooling system. That means that the substantial fraction of fuel energy now wasted as dissipated heat would be saved. Including the savings from other refinements, the methanol car's engine would weigh a third less than conventional ones, which means the whole car could be lighter—Gray and Alson estimate that every pound saved in engine weight means three-quarters of another pound saved by being made unnecessary in the rest of the car. (And, of course, the lighter the car body, the less power the engine needs to drive it, so in turn the engine can be made smaller still.) It takes nearly two gallons of methanol to provide as much energy as one of gasoline, but because the methanol car would be much more efficient the fuel tank could actually be smaller—still another weight savings.

Moreover, the Gray-Alson methanol car would save that serious fraction of the fuel we put into their tanks that our current cars waste in idling.

There are times when not even the most car-loving person loves the car he's in, and that is when he's stuck in a traffic jam. As much as 15% of all the gasoline burned in

America—up to 25% in places like Los Angeles—is burned by cars that aren't moving. That's to the advantage of no one at all, except perhaps to the short-term advantage of the oil industry.

Astonishingly, the Gray-Alson methanol car would never have to idle at all. When the driver stops at a traffic light, or is stuck in a traffic jam, the motor automatically turns itself off. A flywheel continues to spin, so that its stored energy will allow the car to start up again instantly when the driver steps on the accelerator pedal. The savings in fuel—and in pollution—from avoiding idling alone would be immense, particularly in stop-and-go driving.

The Gray-Alson design has a good many other refinements, some of which could well be adapted to gasoline cars. For example, the Gray-Alson car saves fuel in another way by using "dynamic braking"—storing the unwanted energy of motion in the flywheel when the car slows down, so that it can be recovered to accelerate the car again after the stop—instead of wasting it as heat dissipated from the brake linings.

All in all, it sounds as though the Gray-Alson alcohol car, or something very like it, ought to be the car we all drive in the future . . . provided we car buyers are willing to make it so.

There is one major trouble with switching to alcohol fuels. That is the chicken-egg problem: which comes first?

If there were a substantial fleet of methanol-burning automobiles on the road, the oil companies would be quite willing to install a row of alcohol pumps in each filling station to serve them. Contrariwise, if the alcohol filling stations were at every corner there would be nothing to keep the adventurous-minded from being the first on their block to drive a methanol car, or to discourage the car makers from replacing one of their generally indistinguishable models of gasoline-burners with alcohol versions. But which industry can you expect to put up the investment money—$55 billion to get all those pumps and delivery systems installed—to get the conversion started?

There is an interim solution. "Variable-fuel" cars, which can burn either gasoline or alcohol by means of computer-controlled carburetors and other refinements, are on the horizon. They'll cost more. They won't be as efficient . . . but they will be a start.

We mentioned that one of the great causes of automobile pollution is the idling engine in the car that's stuck in traffic. One technological fix for that problem would be the Gray-Alson alcohol-burning car, but there are some useful *social* fixes that could be employed, too.

The roots of the problem of idling cars is clear to every Los Angeles commuter, or businessman trying to catch a plane at Chicago's O'Hare Airport, or tourist trying to drive across the clogged streets of midtown Manhattan:

There are simply too many cars trying to move through too little driving space.

The best a rush-hour driver can hope for is to creep along, stop and go, wasting gas and time. The worst is all too common. That's when things really cramp up, whether it's because of an accident, or road construction, or a strike, or simply an unusual amount of holiday traffic. Then the drivers face gridlock.

That happened in Los Angeles on the San Diego Freeway, one Labor Day weekend, when planes were taking off with half their seats empty, because the passengers who should have filled them were stuck in frozen traffic on the approach roads. It happened again on the same San Diego Freeway in 1986, when a single bad accident tied up the freeway for eight hours. (It wasn't much after that that freeway drivers began shooting at each other.) It happened in London in 1987, when a closed tunnel diverted more traffic into city streets than they could handle and, for seven hours, nothing moved in much of the city—not even Queen Elizabeth's royal Rolls-Royce, with the queen sitting in it along the Thames embankment.

The problem has increased all through the twentieth century. The favorite solution of city planners has been to add more roads, especially expressways with limited access. It still is their favorite solution, in spite of the fact that by now they have all seen that it doesn't work. More roads simply attract more cars. Without exception, every road improvement has increased the number of vehicles using it. Then the new roads become as congested as the old, whereupon newer roads still are built.

Can the problem be attacked from the other end—say, by *removing* some of the cars?

That has been tried in a few cities, and at least proposed, in one form or another, in almost all of them. In the densest parts of most cities curbside parking is discouraged or even prohibited entirely, in order to free up more space for moving vehicles. (But then more cars are attracted again, and anyway that simply creates the new problem of providing off-street parking spaces.)

A better plan is to discourage the private cars from coming into the most congested streets at all, or at least to place limits on their use of them. New York tried that with its creation of "bus lanes" on major avenues, while Los Angeles reserved some freeway lanes for car-pooling.

Because the automobile lobbies have so much political clout in the United States, very little more than that has occurred there. In a few other countries, things have gone a bit farther.

Singapore began charging its drivers for their use of downtown streets with what is called "road pricing" as early as 1975. The driver who wants to take his car downtown in Singapore has to spend a couple of dollars a day to buy a sticker for his windshield, which serves as a sort of passport.

That worked reasonably well in tiny Singapore. It is less effective in larger cities. There it is difficult to enforce, because there are more roads in and out and because it is labor-intensive. Somebody has to look for the stickers, and then ticket the drivers who lack them. Hong Kong tried automating the system with an electronic "sticker" in 1983. In the Hong Kong project, radar-like transmitters along the road scanned each passing car for its electronic signal, somewhat along the lines of the "IFF" electronic identification military aircraft give their allies so they won't be shot down by mistake. Then at the end of the month each car owner was sent an itemized bill for the time they spent in the high-charge district.

The advantage of that was that it was easily handled by computers. The difficulty was that not every driver bothered to install his electronic identifier, and checking up on them was, again, costly in labor. Partly for that reason—and partly because organized car groups there applied almost as much political pressure as American ones do—Hong Kong abandoned the system after a two-year trial.

In England, Norway and the Netherlands projects are under way for more sophisticated control measures. The one in Cambridge, England, attacks the problem of traffic jams directly. What it penalizes the driver for is not simply being in the city center, but specifically allowing himself to be caught in a traffic jam there.

Like Hong Kong's, the Cambridge system requires an electronic "permit" for being in the city center, but the charge is not for a fixed period of time. Instead, a computer in the car makes note of the fact that the engine is running but the car is not moving—as when caught in stalled traffic—and starts a clock running until the car moves again. There is a charge for each minute, prepaid before the car enters the area. If the accumulated charges use up the prepaid amount the computer automatically disables the car. Then the car must be towed away before it can be operated again.

London is planning "electronic toll booths" to divide the center city into regions; each time a vehicle crosses the invisible border between one region and another its computer rings up a charge. This is expected to reduce center-city traffic by a third and, even more important, yield more than half a billion dollars a year that will be used—they say—to improve public transportation.

In principle, the London plan is very promising. It discourages excessive driving by making it expensive, and at the same time it makes the use of public transportation more attractive, and that is the best way to deal with the traffic problem in cities.

Easier and quicker systems that might be inaugurated in U.S. cities include beginning to charge tolls on freeways (sure to present legal problems, as the car lobbyists would instantly take the highway authorities to court), increasing tolls on tollways, bridges and tunnels and taking one lane out of each superhighway for a light rail system.

Environmentally speaking, of course, whatever cures we might impose on the cities would not be enough. There remains the very large problem of cars in the suburbs, taking the children to school, driving to the supermarket or just visiting, as well as the cars which drive cross-country, whether for business or pleasure. The exhaust fumes from a car on an interstate contribute as much pollution to the atmosphere as from one barreling along some ideally clear city street.

So far we've been talking mostly about passenger cars.

The same arguments would naturally apply to such related vehicles as taxis and trucks, but those are only one part of our vast fleet of transportation. People travel in other ways—trains and planes, for instance—and a lot of transportation isn't for people at all, but to carry freight.

Before we get off the subject entirely, let's look at some of those other forms of transportation, starting with—

Surface Freight Carrying

When President Dwight Eisenhower launched his massive plans for building interstate highway systems in the 1950s he probably didn't intend to destroy the American railways.

That was what he did, though. The passenger trains were already threatened, as air travel was beginning to lure away the human travelers who had previously bought tickets on the high-speed special express trains with the glamorous names. Worse still was what the cheap and omnipresent interstates did to the railroads' freight business. After all, passenger traffic was only a fringe benefit for the railroads. It was the millions of tons of freight that paid their bills, and with the new high-speed interstate highways that business began to disappear into the huge new tractor-trailer trucks, some of them now in tandem designs as long as any freight car.

Whether this was a good thing for the country back in the 1950s is still a question that can be argued either way. There is a lot less doubt that it has turned out to be a bad decision for us now. Truck traffic causes most of the wear on the nation's highways. Truckers say that may well be true, but through their high license fees they pay for all necessary repairs, and even for building new roads. The anti-truck people deny that; they say that truck licenses would have to be doubled to pay for what a forty-ton load at seventy miles an hour can do to any highway . . . and certainly the presence of these high-speed juggernauts sharing the highways with the flocks of passenger cars is very bad in another way, because it kills people. Although trucks constitute a minor fraction of the vehicles on the roads, a disproportionate number of traffic deaths come from the people who are unlucky enough to be the ones who are in the cars when there is a truck-car collision.

All that is as it may be. What is an important fact for us environmentalists is that if someone has a ton of cargo to move, it generally produces more pollutants when it is moved by truck than when it is moved by train.

Trying to reverse Eisenhower's decision now, nearly forty years later, will be a hard job. The long-haul trains haven't simply stopped running. They've been erased. Their tracks have been pulled up for scrap, their rights-of-way have often been abandoned. To turn the clock back to rail transport of freight will be dauntingly expensive—especially when we consider all the other urgent demands for capital investment which lie not far ahead.

But it would pay off, if only we could nerve ourselves to do it.

There's an even older transportation network which we could revive now to our advantage—if we still had it. Before the railroads came along the most efficient way of transporting cargo was to put it on horse- or mule-drawn boats and tow the boats along canals. The American canal system flourished for forty years, and it accomplished miracles. It was the canals which pulled together the farms and mines and factories and cities of the United States. They can fairly be said to have played a large part in uniting the United States when, for the first time, cargoes could be shipped by canal from one inland point to another almost as cheaply as sailing ships could carry them along a coast.

Surprisingly, there is an environmental (and maybe even an economic) logic to the idea of reviving the canals, even now.

It isn't just nostalgia. It's true that the idea of floating cargoes along a quiet and pristine waterway through woods and villages and farmlands has a great deal of romantic charm. In fact, it is the sort of thing that many tourists pay heavily to do, when they sign on for cruises along the few surviving canal systems of England and France.

But can it be called in any way *practical*? After all, the principal reason why trains superseded canals is that canals were hopelessly far too slow to compete—three or four miles an hour was a good pace for a canalboat, while even the earliest steam locomotives pulled their trains many times faster. Yet the canals could have a practical value even now, as we approach the twenty-first century. Waterborne transportation is *cheap*; it floats on frictionless surfaces and thus needs less energy to carry goods than almost any alternative. And, as some English researchers have pointed out, in the case of many bulk cargoes travel time is not important. When a load of coal reaches a power plant it isn't burned at once. Instead it is dumped into a huge stack, and there it stays for days or weeks before it is scooped up and burned.

That being true, whether the coal has taken forty-eight hours or two or three weeks to arrive makes no particular difference. Indeed, if canalboats were used to carry such cargoes (not just coal, but ores, grains, almost anything that moves in bulk), the canals themselves would serve as the storage dumps. It would be a fairly simple problem in logistics to schedule delivery times so that the cargo arrived just when it was needed, just as many factories now schedule delivery of parts to arrive just when they are needed, with no inventories of stores kept on hand.

But it can't happen here. Reviving the canals might even be made to work today—at least, to a limited extent—in countries like England or France, where remnants of the old canal systems survive. In the United States, alas, the canals do not exist any more; all that is left of them in most places is a filled-in ditch now turned into a roadway for automobiles—and usually named "Canal Street."

Although it's too late for canals, it may not even now be too late for railroads in America. The Japanese and French have proved that there is a viable place for modern, high-speed rail transport. The Japanese "bullet train" and the French *Trein à Grande Vitesse* have been competing successfully with the short-haul airlines in those countries for years. They will carry you from Tokyo to Osaka, or from Paris to Lyons, cheaper and more comfortably than the airlines. Moreover, for most travelers they will actually do the job faster, because in the train you start from one city center and arrive at another, with no long ride to and from an airport to use up your travel time. The high-speed railroads even make an operating profit while they do it. And they are not the end of the line. The Germans and Japanese have gone far with prototypes of "maglev" trains—using magnetic repulsion to levitate the train above the ground, so that wheel friction is eliminated entirely and airliner speeds can be attained.

How practical maglev trains would be for some future railless "rail" network in America depends on many imponderable factors—economic, environmental and technological. The principal technological question is whether magnets can be produced that are powerful enough and cheap enough to do the job. This in turn probably rests on whether the phenomenon of superconductivity can be tamed and applied to the task; with superconductive power lines feeding those magnets the prospects would seem hopeful, without them the outlook is a good deal less so. A great deal of current research is now going into trying to find the answer to that question.

Then there are a good many other questions: is it still economically possible to get rights-of-way for a major new rail system at all? And—a serious question, to which there is no easy answer—how do you keep people from stealing the large amounts of expensive copper, or other conductors, that would probably have to be left exposed and unprotected along such maglev rail lines?

But even without maglev, there ought to be a future for high-speed rail cargo transportation along the lines of the TGV and the bullet train—possibly cheaper, certainly faster and, above all, less environmentally disastrous than the present fleet of fuel-guzzling, poison-emitting trucks that carry our goods.

Air Transportation

More and more of the travel and cargo movement around the world goes by air. It's fast, and it isn't even terribly expensive—in dollars—any more. Of course, it also carries some pretty high environmental costs. Many householders near a major airport curse the day the deafeningly noisy jet plane was invented, and there are recurring safety worries, especially since President Reagan decimated the ranks of the air-traffic controllers in 1981.

It would not be easy to abandon airplanes entirely. For some purposes it is hard to imagine any satisfactory substitute for travel by air.

It is a lot easier and more practical to think of ways in which we could advantageously replace the fuel-intensive, noisy and inflexible jetliners of the present with something less hostile to the environment. Jet aircraft are monsters of fuel consumption, and a significant source of air pollution.

Worse, the faster they go and the higher they fly, the more dangerous pollutants they release in those high-altitude places where the environment is already seriously threatened: the first studies into possible damage to the ozone layer were initiated because of fears that such supersonic aircraft as the Concorde and many military planes could do what, in the event, was done for them by CFCs. Those studies showed that, actually, the Concorde itself was not likely to do any serious damage to the ozone layer . . . but that faster and bigger supersonics, such as the "Orient Express" aerospace plane proposed by President Reagan, could indeed be an additional menace to our already threatened ozone.

One way to make jets less poisonous to the environment would be to burn something other than petroleum products in them, just as with cars.

That's not out of the question. Alcohol might be substituted. Surprisingly, so might hydrogen—indeed, at least one large hydrogen-fueled jet has already flown when, in April, 1986, the Soviets made the first such successful flight with a hydrogen-modified Tupolev 155. Hydrogen might be better as a fuel for aircraft than for cars, in fact. Because hydrogen requires complex and heavy storage, it would work best with large vehicles. With aircraft it would work best of all; indeed, an airplane designed from scratch for hydrogen fuel would have a large competitive advantage in the fact that it would weigh 10% less than a conventionally fueled plane at takeoff.

But perhaps the jet plane itself can be replaced. The jet has one great advantage. It flies faster than the propeller-driven aircraft it has taken over from.

Curiously, though, there are ways in which that advantage can be made to disappear. Some English studies have produced a design for a propeller-driven plane which would cruise at nearly jet speed—and would not have to fly as high.

This is a very big advantage, because a lot of a jet's flight time, and even more of its fuel consumption, is used in reaching its cruising altitude and coming down from it again at the end of the trip. The result of the lower cruising altitude would be that in flights of a thousand miles or so—say, New York City to Miami, or Atlanta to Chicago—if one of these planes took off at the same time as a conventional jet it would actually beat the jet to its destination. More than that. Since propeller-driven planes typically can land and take off in shorter distances than jets, airports can be located closer to the cities they serve, thus making additional savings in time and fuel for the prop planes.

Finally, there's no law that says all air travel has to be in any kind of a plane. We could give lighter-than-air craft another trial.

The Germans did operate their Zeppelins on transatlantic routes with considerable success for some years. The cruising speed of the Zeppelins was quite slow—about the same as a car—but still faster than any ship in crossing an ocean; the comfort was considerable, with passengers enchanted with the view of the seas and landscapes they passed over at low altitude; and, since no energy from the motors was required simply to keep the vessel airborne, their fuel cost was relatively low. The Germans were widely envied for their lighter-than-air vessels . . . at least, until the *Hindenburg* crashed and burned in Lakewood, New Jersey, in the late 1930s.

That put an end to commercial lighter-than-air transportation. It needn't have, though. What doomed the *Hindenburg* was not a built-in flaw intrinsic to all lighter-than-air craft, but only that its lifting gas was combustible hydrogen instead of inert helium. The reason the Germans used hydrogen was that they had no alternative. They had no source of helium of their own. The United States, the world's biggest supplier of helium, had plenty; but in those tense times just before World War II the United States would not sell to the Germans a commodity which might some day be turned against America as a weapon of war.

None of those problems hold true any more. Helium is produced in quantity, in the process of producing natural gas; in fact much of it simply escapes to the air and is wasted. There are designs now on the books for helium lighter-than-air craft which could do everything the old German Zeppelins did, but better; could carry cargo cheaply and, if not more quickly than by jet, certainly more quickly than any surface transportation. And the savings in fuel would be immense, for the reason mentioned already: the engines in a lighter-than-air vessel need only propel the craft through the air; they don't have to waste the major fraction of their output simply to keep it in the air, as all heavier-than-air planes do.

And saving fuel, of course, is a wonderful way to reduce pollution drastically.

Before all these technological improvements show up, how do we cut down on the poisons of vehicle pollution now?

We can try the same trick we proposed for fuel-burning factories and power plants: internalizing their external costs by means of a tax. We would have a good use for any tax money that could be collected that way. We could spend some of it to restore the research funds the Reagan administration cut, hiring the scientists and engineers who could make all these prospects real.

How much should the tax be? An added fifty cents a gallon would be a good start. It might help to discourage excessive fuel burning. It certainly could produce ample funds for a serious research program, since in round numbers each added penny in gasoline tax brings in about a billion dollars in revenue.

Even with a fifty-cent increase U.S. gasoline would still be far cheaper than in most of the rest of the world; in most of Western Europe it costs twice as much as here.

Which may be one of our problems. If we had grown up in a country that paid a more nearly appropriate price for gasoline we might be more careful with it. As Tom Paine said, "What we obtain too cheap we esteem too lightly; it is dearness only that gives everything its value." Dearer gasoline might make us rethink our feelings about those poison-gas factories we so casually drive to the corner convenience store.

17

Fixing the Home and Farm

If we ran all our cars and factories on electricity and generated all our electricity from sunlight we still would not have cured all our environmental ills. Our homes and our farms would still need a lot of work.

Start with the home we live in, the place where we spend most of our time. There's a great deal that can be done to make it more environmentally benign.

Gadgets and Appliances

Every American home is filled with power machines and appliances, everything from electric pencil sharpeners to furnaces. The reason we buy these things is that they are supposed to contribute to our comfort and convenience. Sometimes they do . . . though at some other times the actual benefits they produce are hard to find. (With the possible exception of a few people with severe arthritis, does anyone really *need* an electric can-opener? Or a noisy, ponderous, gasoline-burning leaf-blower?) What all our gadgets do have in common is that collectively they use a great deal of energy, and therefore produce a lot of pollution.

How much pollution is "a lot"? Maybe more than you might think. The appliances in an average American home use so much electricity that the fossil-fuel plant that supplies them produces *five tons* of carbon dioxide a year to make them go.

It's easy to find gadgets on sale that claim to protect us *against* pollution. There are air filters and electrostatic ion generators, which promise to take the pollution out of our indoor air, and water filters which fit over the kitchen taps and are supposed to remove all unhealthful pollutants and undesirable tastes from what we drink and cook with. Some of them work. Most don't, being either useless to begin with or rapidly deteriorating to become so unless they are carefully maintained. Appliances which *reduce* pollution are harder to find.

As always, the quickest way to cut down on the pollution we cause with our household appliances is to use less energy to run them. A four-slice toaster uses more electricity than a two-stack, and if you are only toasting one or two slices at a time you are wasting most of that extra electricity. That means, of course, that you are forcing your local power plant to generate that much extra pollution. (You may still want to be able to toast four slices at a time for Sunday breakfasts, but there's a solution to that. Buy a four-slice toaster with two controls.) When you shop for a new refrigerator compare models. You'll likely find two refrigerators side by side, the same size, holding the same quantity of food, keeping the same interior temperature, but one may be constructed with extra insulation. That one will cost more to purchase—somewhere around an extra $100—but it will also use so much less electricity to run that over its 15-year service life you will save several times the added cost in electricity bills.

The way you use your appliances makes a big difference in how much energy they use. Don't run your dishwasher until it has a full load; you'll save power (burning less polluting fossil fuel in the utility power plant), as well as detergent (reducing the pollution of wastewater) and water itself (in increasingly short supply in most areas). When your laundry dryer finishes its cycle, take the clothes out; if you delay, most modern dryers will patiently start up again with a short heating and tumbling cycle to keep the clothes from wrinkling, and they will go on repeating the cycle over and over until you get around to removing the laundry—burning more wasted electricity every time they recycle.

You might even consider going back to muscle power instead of gasoline or electricity for some of your household machinery. A hand whisk makes better whipped cream than an electric beater. If you're reasonably healthy and your backyard is less than a quarter acre, do you really need a power mower?

It isn't just the fuel your power equipment draws in operation that adds to the pollution, of course. Power equipment generally is more expensive and more energy-intensive to manufacture than the hand-operated equivalent—and thus creates more pollution even before you ever take it home from the store.

Lighting

The fewer watts you draw for electric light, the less oil or coal your local power station has to burn to generate electricity.

Of course, you don't want to live in the dark. But you can have just as much light when and where you need it and still reduce your electric bill (and the power requirements for your local public utility) significantly.

The amount of light given off by a bulb is measured in "lumens." Ordinary incandescent bulbs, the kind you grew up with, produce around 14 lumens per watt of electricity. You can do a lot better than that. The newer tungsten-halogen bulbs get about 20 lumens for the same watt. There are disadvantages to the halogen lights. They're expensive (but they last three or four times as long as ordinary incandescents before they burn out). They are delicate—it's best not even to leave fingerprints on them, because the oils in your fingers may produce local hot spots that will damage the bulb—and they emit significant amounts of ultraviolet light, which can harm the eyes in excess, so it's not a good idea to look at them directly. (There's not enough to give you a sunburn or any of the other perils of solar ultraviolet.) Finally, the halogen lamps run very hot; they should only be used in fixtures which are designed for them, with plenty of ventilation to carry the excess heat away . . . but when you have factored in all the disadvantages, the balance sheet still shows their one great virtue: they are far better at converting electricity into illumination than ordinary incandescents, and so they do not cause as much energy-related pollution of the environment.

There are ways of improving the performance of incandescent bulbs. The electrical current passing through its filament produces two kinds of radiation: visible light, which you want to read by, and infrared, which you don't want because it's wasted as heat. Two new bulb designs help save that wasted infrared. In a new General Electric design the inside of the bulb is coated with a film that reflects infrared (but not visible light) back inside to make the filament hotter, and thus to burn more brightly. A bulb patented by Nelson Waterbury contains two filaments. One runs off your house current in the usual way. The other one is powered by a tiny photovoltaic cell sensitive to the wasted infrared.

Fluorescents have been around for a long time and they are another alternative. Fluorescent lights use substantially less energy than the old incandescent bulb—they produce from 40 to 80 lumens per watt, several times as much as the incandescent—and the new "surface-wave" fluorescent designs are better still.

Compact fluorescents have just come onto the market. They cost much more than incandescents—currently $15–$18 for an 18-watt bulb—and, although that bulb gives light equivalent to a 75-watt incandescent, they may not look like a real bargain at first—especially if, as is true with some public utilities, your local power company will provide you with free incandescents every month.

If you are a dedicated environmentalist you might grit your teeth and accept the cash loss for the sake of keeping all that pollution out of the air. Actually you don't have to. The compact fluorescents will give you a *profit*, because there's more of that bad bookkeeping here. The compact fluorescent lasts ten times as long as the incandescent it replaces, and each one of those incandescents will burn about $5 worth of electricity in its life. (Now you know why your friendly utility is so willing to give them away.) So—depending on what the utility rates are in your neighborhood—you will save at least twice the cost of the compact fluorescent over its lifetime, as well as saving the trouble of replacing a bulb ten times. You also add less heat to your house in the summer . . . and, of course, you help in the things this book is about, because you cut down your personal contribution by a substantial amount, since the generation of power for a single 75-watt bulb produces 200 pounds of carbon dioxide a year.

(Actually, the compact fluorescent isn't *quite* as good as we've just said. For technical reasons, the apparent draw of some compact fluorescents—the amount of energy you pay the power company for—is somewhat less than its "real" draw, the amount of energy the power company has to generate to make it run. The difference, though, is small. Some compact fluorescents take a moment or two to light up after you turn the switch, too, which may be disconcerting at first.)

You might not want to replace all of your ordinary incandescents with halogens or fluorescents, but in the proper applications they can be very useful.

Many householders are buying "long-life" incandescents to save the bother of replacing burned-out bulbs as often. They will do that quite well; however, they will also burn more energy per lumen than the conventional kind—that is why they last longer. It's a good idea to reserve the long-life bulbs for places where replacing a burnout is particularly difficult; otherwise you are paying more in electric bills, and in pollution.

For incandescent lights of all kinds (but not for fluorescents) a dimmer switch can save a lot of energy: you may need all six lights in an overhead lighting fixture burning at full brightness when you are cleaning or reading, but only a fraction of that illumination while you are watching television. With the dimmer switch you have total control of how many watts you burn.

It's worth remembering that there is a very elementary way of saving electricity in your lighting system, of course. It is the one our parents tried to teach us long ago: you simply turn off the lights you don't need.

If it's too much trouble to remember to do that yourself, you can have it done automatically. In many parts of the world (but seldom in America) most apartment hallways, for example, are equipped with time switches: the tenant turns them on as he enters the hallway, and a few minutes later the lights automatically go off. That is pretty crude stuff, but better technology is becoming available in the form of "smart" lighting switches. These sense human presence so that the lights go on as soon as anyone enters a room and go off again when the room is empty.

Most of these improved lighting systems are already "off the shelf." You don't have to wait for inventors to come up with something better, you can go out and buy them today. And you'll not only do your bit to cut down on pollution, you'll actually save money on your electric bills. According to Amory Lovins, improved lighting efficiency could cut America's consumption of electricity by 25%— not only saving the environment all that pollution but saving the country more than $30 billion in utility bills every year.

Heating and Cooling

The major use of energy, and thus the major source of pollution, in the American home is its heating and cooling system.

Your home heating system produces more carbon dioxide than even your appliances—seven tons of it a year.

The quickest fix, of course, is to be more tolerant of minor temperature changes—to turn the air conditioner thermostat up a few degrees in the summer, and put on a sweater in the winter. That would save a good deal of energy in itself, but there are even more effective ways—some that can be "retrofitted" to most existing homes, others which require longer-range planning.

In the long range, for instance, there is a lot of free heat around which could be used to heat homes at no fuel cost at all.

That is the waste heat from industry and power plants.

Industrial-strength heat is *hot*—hot enough to boil water into steam and sometimes, as process heat in steel mills and chemical plants, a lot hotter. The low-temperature heat that is left over when the high heat has done its job—the kind that you want in your home—is a drag on the market in industry. It is something the plant owner has to pay to get rid of. Generally it is dissipated through cooling towers, uselessly warming the air around the plant, or by pumping water from the nearest river or lake, thus warming them and as often as not making life difficult for their fish and other organisms.

But that heat needn't be wasted. It can be piped to nearby homes and offices, and in fact this is done, very successfully, in some northern areas. The surplus heat doesn't have to go just to residences. In a number of places in northern Europe, for instance, they have built large-scale pumps and pipelines so that the waste industrial heat goes to warm huge greenhouses, where "out of season" fruits and vegetables are grown for the local markets.

(Remember that most "waste" is only something useful that's in the wrong place—so all this secondary use of industrial heat does is take the heat from where it is a nuisance and apply it where it does good.)

Then there is the free gift of solar heat, which is available everywhere during most daylight hours and basically free.

Using solar heat in a home requires building a heat collector on the roof—those dark rectangles one sees on a few homes in most areas—and then pumping that heat, as hot air or hot water, around the home. This can be done fairly easily as a retrofit on most existing homes. The capital cost is high, but the fuel savings are considerable.

Of course, if you install the solar heating in a new home while it is being constructed it's a lot cheaper, and also a lot more efficient. Also of course, if you use such devices on a larger scale than the one-family house you can do better still.

On public buildings, for instance, solar heaters can not only use transient heat, they can trap it in the summer and store it up to be withdrawn again when it is needed in the winter. The University of Massachusetts at Amherst has a 12,000-seat sports arena designed to be heated almost entirely in that way. Solar-heated fluids are piped through a network of pipes through a huge bed of underground clay in the summer. The clay warms all summer long; then, when cold weather arrives, the same pipes carry its heat back to the arena. The capital cost is high, but it is designed to pay that back through fuel savings in seven years—thereafter the heat is essentially free.

(The same summer-to-winter trick can even be done in reverse, by storing up winter cold and extracting it in the summer to save on air-conditioning, for instance. The key is the "ice pond," allowed to freeze in winter and providing a sink for heat in the summer. The basic idea of the ice pond has been around for a long time, but Ted Taylor, with input from a number of others including Freeman J. Dyson, has been developing new high-tech versions which offer considerable energy savings for both homes and industries.)

More efficient home furnaces can save a good deal of fuel. Most newer American homes do not have steam or hot-water heating systems, but are heated by forced-draft hot air: the furnace burns a fuel, usually oil or natural gas, which heats a large piece of metal; a fan blows air across the piece of metal and conducts it to the rooms of the house.

Unfortunately, a good deal of the heat goes right up the chimney and is wasted. New burner designs save fuel by burning the fuel more frequently but for shorter periods of time. Some designs have thermostat-controlled fans in every room, monitoring the flow of hot air on an individual basis, so that rooms not in use can be left cooler until needed.

Then there is insulation.

In winter we want our homes warmer than the outside air, in summer we want them cooler. Year around, the less interaction we allow between inside and outside the home, the less energy we have to use to keep them that way. As our high-school science courses teach us, there are three main mechanisms for heat to cross the barrier of our house walls: radiation, convection and conduction.

Radiation is just what it sounds like: radiant heat from the sun enters our homes, through our windows, and exits in the same way in the winter. The bigger the area of glass, the more heat is gained or lost; picture windows are major enemies of energy saving, since they transmit a lot of it. Double-glazing helps; so do shutters, blinds or curtains.

Convection is the carrying away of heat by movement of air. Leaky windows allow a lot of heat to enter or be lost; the solution is tightly fitting windows, preferably helped out by puttying or taping around the edges.

With normal construction, that sort of leak means that the entire interior air is exchanged with outside air about ten times an hour—and all that exchanged air has to be heated or cooled to the comfort range. Just stopping the major leaks can cut that down to one change per hour. Replacing standard windows with state-of-the-art ones, coupled with extra insulation, can cut these convective losses so far that many houses would hardly need a central heating system at all; the waste heat from the hot-water heater will keep them comfortable through the winter—even in areas like Massachusetts.

The tighter the house is sealed, the less heat transfer will occur through convection. But there too there is no gain without pain; the pain in this case means that the household air isn't as fresh as it might be. (On the other hand, if the outside air is sufficiently polluted you may not want it in your house anyway.)

Most of us have become aware of another threat that may affect some well-sealed homes: radon may accumulate in the household air. Radon is a heavy gas which comes from the natural decay of such radioactive elements as uranium in the soil. Radon is damaging to health, and has been called the second most frequent cause of lung cancer after smoking. Radon is emitted naturally from some kinds of rock, like granite, and especially from the waste from industry which uses uranium or other radioactive elements; years ago, that waste was eagerly sought for fill by builders, and a good many houses in some areas are built on this mildly radioactive waste.

But only a small minority of homes have a radon problem. If you are concerned, the best first step is to have your home tested to see if the problem exists for you. (The test is cheap and easy.) If you find no significant amount of radon, then opening a window now and then will clear out most of the other domestic gases. If you do have a serious amount of radon (which is to say, a reading of more than 50 millisieverts per year), the best solution isn't to ventilate your house more, but to prevent the radon from entering in the first place. That means installing pipes and a fan under the foundation of the house to suck the radon away before it comes into the house at all. Doing that is not cheap—anywhere from a few hundred dollars to several thousand—but it is still cheaper than lung cancer. (Of course, the best solution of all is to avoid building homes on radon-emitting foundations in the first place.)

Conduction is heat transfer that occurs when a cold object actually touches a warmer one. When your outside walls are under a blazing July sun they get hot, and that heat is conducted to the inside; in the winter, the transfer is in the other direction. What controls the amount of heat gained or lost through conduction is the amount of insulation in the outside walls.

Rick Bevington and Arthur H. Rosenfeld made a study of the effectiveness of insulation. In their report (published in *Scientific American*) they compared two new Chicago-area houses, one equipped with the standard 3½ inches of insulation, the other with 6 inches. The extra cost for the thicker insulation was $300. The heating costs thereafter for the standard home were $200 a year; for the better-insulated one, only $80. That's a return on investment of 40% a year . . . not counting the environmental saving of those several tons of carbon dioxide each year.

The insulation is usually some sort of loose material, with a low conductivity of heat, such as glass fiber, placed inside the outer walls and in the attic or roof. It is frequently possible to add insulation to an existing house (by blowing the insulating material into the walls and laying batts of it on the attic floor), but this is much more effectively done when the house is being built.

But there is a more fundamental change we could make in our housing. If we want to attain real heating and cooling efficiency, the standard American single-family home is simply a hopelessly bad idea in the first place.

To see why that is so, think of a house as a cube.

A cube has six faces. The more faces that are exposed to the air and sun the more exposed area there is for heat to travel through. A single-family house has five exposed faces—that is, the roof and the four sides. The floor's exposure is unimportant since it is in contact with the earth, and thus is not exposed to radiation or convection losses, while the conductive losses are minimal in the basement since the ground does not get as hot in summer or cold in winter as the outside air.

If you cut down the number of exposed faces, you cut down the heat loss accordingly. The "semi-detached" house, in which two houses share a common wall, exposes one less face of the cube; therefore it automatically reduces heat losses by

20%. The apartment is more efficient still. In a four-story apartment building, most rooms expose only one or two faces to the elements (the top-floor apartments expose two or three) and heating and cooling them is correspondingly cheaper.

For those who do not want to give up the idea of a home that is all their own, there remains one rather radical alternative to the conventional split-level or ranch house. Why not build your next home underground?

"Cave-dwelling" has an unattractive sound to many people. It's true that if you build underground you must sacrifice a view and you can't open a window for fresh air (but how often do people open their windows in homes with central air-conditioning anyway?). Problems may arise with underground springs or a high water table; the site has to be carefully chosen, and drainage provided when necessary. There may also be difficulties with present local building codes, which were not usually intended to take care of radically different kinds of home construction.

You will also need artificial lighting day and night (though a certain amount of natural daylight can be "piped" to the interior of the house from skylights on the surface). Nevertheless, the ecological and economic advantages are great. The underground house is a nearly perfect design from a heat-exchange standpoint. Most of it isn't exposed to the weather at all, since it is surrounded by rock and soil—very good insulators—and therefore heating and cooling are minimized all year around. "Terratecture"—the designing of such underground homes—is beginning to be employed in some areas already, and its future is bright.

Terratecture doesn't just save energy, it can do more. When the architect Malcolm Wells decided to build himself an underground office in Cherry Hill, N.J., he scooped out a hole in the ground, covered it with a waterproofed concrete roof and covered that in turn with the subsoil he had excavated. Subsoil is dead soil. To make it come to life, Wells trucked in dead leaves from a nearby city dump and spread them on top. That was all he did. The next year weeds sprang up; within five years he had essentially climax vegetation over his office, without seeding, buying topsoil or using chemical pesticides or fertilizer.

In London's Kew Gardens, the Sir Joseph Banks building that holds the bulk of their plant collections is largely underground. Heated by heat pumps, with shafts going down to groundwater to exchange heat summer and winter, it has been called "the most energy-efficient public building in Europe."

All that we've said so far, of course, barely skims the surface of what can be done around your house to help save the planet we live on. We've been talking about saving energy, but homes are great water wasters, too, and there are many simple things you can do to cut down on the depletion of our water resources.

You can start by checking for leaks in your plumbing. The way you do that is to turn everything off, then go look at your water meter to see if it's still turning. If it is, there's a leak you don't know about. One of the most likely places for such a thing is in the toilets; pour a little food color in the tank and see if it shows up in the bowl without flushing. Or go down in the basement and put your ear to the pipes; often you can hear the faint hum of leaking water when you can't see it.

If you want to go farther, and you would like to keep your

lawn and garden green without wastage, think about gray-water.

"Graywater" is the stuff that comes out of the drains of your bathtub, shower, washstands, washing machine and (unless you have a dispose-all garbage grinder) kitchen sink—out of every bit of plumbing you own, in fact, except the toilets. (You don't want to do anything with the water from your toilets except get rid of it.)

You don't want to drink graywater. On the other hand, you might be able to use it for hosing down your driveway or washing your car, and you surely can use it well for watering your lawn. There graywater isn't merely adequate. In most cases it's actually better than your drinking water, because of the detergent residues it contains; they turn into fertilizer when they hit your plants and, although the use of fertilizer carries its own environmental penalties, in this case you are merely making use of the chemicals that would in many cases go into the environment uselessly anyway.

To use your graywater, you need to change the drains on some of your household fixtures and install a holding tank and a pump to put the water where you want it. This costs money, but probably only in the hundreds, rather than the thousands, of dollars. Still, it can save money for you on your water bills (you won't have to pay again for the graywater you use) and if you ever face severe drought restrictions it can save your whole garden.

However, you must check with the local authorities to find out if graywater use is legal. Some communities restrict it on health grounds, though that's not a problem if you do it right. Some ban it entirely, mostly because it provides a good chance for cheating in droughts; some people might be tempted to turn on their bathtub faucets and let them run for an hour or two to replenish their "graywater." (Yes, there are a few people like that.)

You might want to take a critical look at your kitchen stove. If it has a pilot light, you are burning a lot of natural gas you needn't—electronic ignition (or matches!) could be a big energy saving. (Assuming you have matches. In parts of the USSR, where shortages of everything have plagued them, they didn't, and so many households left their gas stoves burning all day.)

Above all, when it's time for you to move you should think of all these things in deciding on your new house, especially if you're building from scratch. It's certainly possible to make great improvements in an existing home—but much easier and cheaper to build them in the first place.

The Farm

It's a fact worth remembering that American agriculture is one of the great success stories of human history. We eat well, and the reason is that we grow more food, with the labor of fewer people, than anyone has ever done before in the history of the world.

But that success has come at a cost to our natural resources and even to the quality of some of the food itself. The environmental bills are all coming due, in the form of depleted water resources, deteriorated soil, or even the destruction of soil through erosion and damage to the environment caused by the use of agricultural chemicals.

There are environmentally sound alternatives to most of our damaging farming practices. Here are some of them:

Conserving Water Resources

To see what water means for American agriculture, we can look at the 1988 drought. Over much of the farmland of the United States in that year rainfall was scarce and what there was of it didn't come at the right times. Its lack cut U.S. wheat production by 40%. For one of the few times in recent history, in 1988 the United States consumed more grain than it produced.

Yet our farmers continue to waste water in incredible amounts. A case in point is the California habit of pasturing cattle on water-intensive irrigated lands, to which we've already alluded—over four million acre-feet of water are wasted each year, just for that. It would be far better to pasture milk cows in some wetter region outside the state, importing the milk and meat needed, and use those lands for some more water-efficient crop. Such crops as cotton, rice and alfalfa also require a great deal of water, and yet they too are grown all over the southwest. If some of those pastures and water-intensive farms were converted to uses which require less water—for instance, to citrus groves—the cash return would be as good, and the water demand would be cut drastically.

Most California irrigation is in the least efficient forms, too: sprinklers rain on everything below them, or ditches are filled with water and allowed to soak into the land. (The ancient Babylonians did as well as that thousands of years ago.) As the Israelis have demonstrated, trickle irrigation can save as much as 90% of water for suitable crops.

Don't American farmers know about these better systems?

Of course they do. The reason they don't change to saner practices is simply that the existing water laws, particularly in the southwest, penalize them if they do. Irrigation water has been made so artificially cheap that it simply doesn't pay the farmer to try to save it. Worse, the rule for water is "Use it or lose it." If a few farmers are willing to spend the money to adopt water-conservation methods what results is that their water allowance is then permanently reduced by whatever amount they have saved. They've lost the water and, since the savings in water bills is less than the cost of the systems they've installed, they've lost large sums of out-of-pocket money as well.

It's important to remember that irrigation is not an unmixed blessing. Rainwater is pretty pure (or would be if we did not contaminate it with acids and other chemicals). Irrigation water is not. It contains trace amounts, sometimes much more than trace amounts, of dissolved salts. When it evaporates it leaves the salts behind, and ultimately the chemicals irrigation leaves behind will poison the soil.

It is possible to slow down the poisoning by installing a separate system of (expensive) drains to remove the water from below the crops, but there is no way to halt it completely. Sufficient quantities of natural rain will ultimately wash the salts away . . . but there seldom are sufficient quantities of natural rain, because if there were who would irrigate in the first place?

The only other alternative is to find salt-tolerant crops. Such plants are called "halophytes"; they can thrive on brackish or even moderately salty water. Some of these traits can be developed in crop plants—cotton and tomatoes in Israel, animal fodder in Pakistan, and an oilseed called salicornia in Mexico. There are natural halophytes which could be (but seldom are) grown as food crops, too, such as the Australian quandong, the fruit of which resembles cherries. This is not only the probable future for most irrigation farmers, but a serious present concern for those who are farming on the millions of acres all over the world which have already been damaged by ill-considered irrigation projects. (As well as for those whose underground irrigation water supplies have been damaged by incursions of sea water.)

Farmers have another responsibility, largely neglected, which is to do as little damage as possible to the water they use.

Sooner or later a good deal of runoff from that irrigation water makes its way back into the rivers or the underground aquifers. Unfortunately, that water is often seriously polluted by the time it gets there. Contaminants of agricultural runoff include fertilizer (which leads to eutrophication of lakes), organic waste (largely in the form of excrement from stock herds) and toxic chemicals used to control weeds and insect pests.

Fertilizer

Farmers fertilize their crops to make poor soils grow or to restore soils which have been over-farmed and worn out. The principal element in most fertilizers is nitrogen. There's plenty of pure elemental nitrogen in the air, but that doesn't do most crop plants any good. They can't fix it for themselves, and so it must be supplied to them in the form of assimilable nitrogen-containing compounds. That can be done chemically—that's how synthetic fertilizers are made—but manufacturing chemical fertilizers is a fairly messy industrial process, and organic farming enthusiasts claim the resulting crops are not as good for you as those grown with natural fertilizer.

For that matter, there is some question about whether *any* fertilized crops are as good for you as naturally grown ones. A new element in the discussion was added quite recently when experiments conducted by the U.S. Department of Agriculture in Beltsville, Maryland, showed that leafy vegetables grown with any kind of fertilizer contain only about two-thirds as much vitamin C per ounce as those grown without fertilizer. It makes no difference whether the fertilizer is organic or manufactured. So the cabbages in your supermarket may be bigger—but you may have to eat more of them to get the necessary vitamins in your diet. (The scientific jury is still out on these questions, however.)

Still, there are other natural sources of nitrogen.

One of them is crop rotation—planting lupins, clover or legumes like soybeans, which actually add nitrogen to the soil through the rhizobia bacteria associated with their roots. These bacteria "infect" the plants. That sounds bad, but in this case what the "disease" does is only to supply to the plants the nitrogen from the air which they can't fix themselves. What fixes the nitrogen is an enzyme, nitrogenase, and what produces that particular enzyme in the nitrogen-fixing organism is a group of genes collectively called *NIF* (for *NItrogen Fixing*).

Now that scientists understand that much about the process, they are trying to take the next step. That will be to splice these *NIF* genes into common crop plants. When (and if) they succeed, the need for any kind of fertilizers will be drastically reduced.

It is also possible, of course, to fertilize our crops the way our forefathers did, with the good old animal (or sometimes even human) manure . . . which brings us to the next environmental agricultural problem:

Organic Waste

Our domestic livestock produces several hundred pounds of manure per head each year. When most animals were raised on family farms disposing of the manure was not a great problem; typically it was stacked and allowed to rot for a while, then spread on the croplands.

Factory farming has changed all that. Now we concentrate vast herds of animals in small areas. Chickens are battery grown, two at a time in tiny cages; pigs for pork and calves for veal almost as confined; steers may start on the open range, but they are fed up to market weight in densely crowded feedlots, on farm-grown grain.

This produces other problems in addition to those concentrated doses of excrement. We know that range-grown beef is better for our health (less fat, fewer heart attacks), and that there are ecological, and maybe moral, objections to devoting a third of our grain production to animal feed instead of feeding many more people with it directly. But the problem we are now discussing is the excrement. Pig manure is accumulating in such vast quantities that some of the world—Holland is one example—is simply running out of ways of disposing of it in what is, after all, a rather small country.

It's not just pigs, either. There are seven and a half Dutch cows, pigs, ducks and chickens for every Dutch human and all of them excrete—80 million tons of animal manure every year, and not enough cultivated fields in the whole country to spread it all on. Most Dutch farmland is so well fertilized already that the excess phosphorus, nitrogen and ammonia simply washes away into the waterways, which bloom with algae as a result. The situation is not much better in parts of Italy and France, or even in some parts of Taiwan.

As we saw earlier, one solution is to digest the manure for methane and to use the residue as compost. Another would be to eliminate the feedlots and eat beef derived from range-fed cattle—after all, one of the great advantages of a steer is that it can convert inedible grass into edible food. It might even be a good idea to eat less meat.

It might also be a good idea to use our cattle to *get rid* of some kinds of wastes.

A cow is a wonderful machine for converting an otherwise useless resource that is produced by nature in great volume—namely grass—into high quality food—namely milk and beef. Human beings can't live on grass, because too much of it is indigestible cellulose. Cows can. They have better digestive systems, aided by bacteria that convert the cellulose into sugars in their stomachs.

It turns out they can do the same trick with some other low-valued resources that we have huge supplies of as well.

It's customary to spread straw in cow barns, for the comfort of the cows and to soak up their excretions. Cows normally eat some of the straw, and their microorganisms turn it, too, into food. In England, some farmers found out that waste newspapers were cheaper than straw, and when they began using them they discovered that the cows ate the old papers, too.

That's not really surprising. Cows will eat all sorts of

things, some of them very unattractive. Farmers in the United States have added such things to cow fodder as cement dust, and even the cows' own dried excrement, without doing any apparent harm to milk or meat. A diet including yesterday's papers doesn't seem to hurt them, either.

All in all, farmers can do much for the environment. Indeed, many of them do. In 1980 about one per cent of American farmers were practicing sustainable agriculture. Their crop yields were generally somewhat lower—but so were their production costs, because of savings in purchased chemicals, and so the profitability was pretty much the same. And the environment much safer.

Even fish have begun to damage the environment with their excrement. Salmon farming in the Puget Sound area, for instance, is a thriving business. The fish are kept in two-acre floating pens, fifty to a hundred thousand of them in one pen. Certainly fish farming is a good way to go (especially considering the ways in which we have depleted some of the free fish populations, and made others inedible by water pollution), but the fish, too, excrete. In Puget Sound, each fish-farming pen drops some fifty tons a year of excrement and uneaten food to the bottom of the Sound, which deplete oxygen and cause algae blooming—and the algae not only kill wild organisms in the water but threaten the health of the salmon themselves. Moreover, the fish food is generally laced with antibiotics, which encourages resistant strains of disease organisms to grow in the water. The disease organisms may not stay there. With only a little bad luck, they can easily wind up in people.

Fortunately, the fish problem is fairly easily solved. All it would take to clear up the situation in Puget Sound would be to locate the pens in deeper water, with stronger currents to carry away and dilute the detritus . . . an easier solution, at least, as long as the number of such pens is still relatively small.

Pesticides and Weed-Killers

We've already mentioned that, although American farmers' use of pesticides has increased nearly tenfold in the past half century, at the same time the percentage of crops lost to insects has nearly doubled. That's not all. In some cases, overuse of a pesticide has brought about the development of new strains of the pest which are no longer controlled by that chemical. In others, pesticides which were applied to kill a specific pest have killed that pest's natural enemies as well, thus allowing even more of the pests to survive.

There are better ways, biological ones. More than half a century ago the U.S. government began importing the preying mantis to America in an effort to control some insects; other such predators can be imported and used. Some pest insects recognize each other (especially for sexual attraction) by airborne chemicals called "pheromones"; if these chemicals can be identified and synthesized they can then be used to lure the insects into a trap.

Chemical weed-killers are expensive and leave residues which find their way into water (not to mention their effects on wildlife). Curiously, some crops—for example, sorghum, rye, alfalfa and cucumbers—are natural weed-killers. They have the power to inhibit weed growth. They don't *prevent* weeds, but they do significantly reduce the number of weeds by from a quarter to a half.

Moreover, they don't merely protect themselves. They appear to leave chemicals in the soil which help protect the next crop—no matter what the next crop is.

There are other ways of discouraging insect pests. One good one would be to cut back on our present practice of providing them with nearly ideal conditions for multiplying, by presenting them with vast stretches of farmland containing their favorite foods. This is called:

Monoculture

There aren't very many omnivorous insects. Most of the ones that become agricultural pests strictly limit their diet to a particular species of plant, and often to only a small number of strains of that species.

Before farming became "scientific," most farmers grew the kinds of crops that did best in their particular plots of soil. They saved the best of each crop for seed, and what was best for any individual farmer often turned out to be a strain different from the one that grew on his neighbors' lands. That had benefits for pest control. An insect which thrived on one particular farm might starve on the one across the valley; and if some farms were hard-hit by nematodes or borers, there were many others in the same region which were not affected at all.

Now that crop diversity is gone. Most current farmers buy their seeds instead of raising them, and over large areas they all buy seeds of the same race. The result is that a pest which might once have done only spotty local damage can now devastate a whole countryside.

Moreover, different strains of the same species may have

special qualities not shared by others. In Peru, where the potato began, the old farmers cultivated several hundred different potato strains, with distinct differences in resistance to particular pests as well as differences in flavor and growing qualities. Manioc, a staple starch eaten by hundreds of millions of people, comes in at least 200 separate races, each with special characteristics of its own—one is better at making flour than another, another better for gum; some harvest earlier, some grow taller, the leaves of some are edible themselves.

But all those variant races are disappearing as farmers, worldwide, are learning to buy commercial seed strains and monocultures are taking over. There are certainly advantages to the commercial strains—hybrid corn far outperforms most natural varieties, as do the grain varieties of the Green Revolution. But there are other advantages to the disappearing strains which are lost. Even the ones which can't compete in themselves contain genes which may usefully be crossbred into farm races—if we preserve them.

Conserving Soil

The American midwest is losing its soil faster now than it was in the famous Dust Bowl years of the 1930s. The loss gets less publicity now, because it doesn't advertise itself as conspicuously: the dust storms of the '80s modestly kept themselves near home instead of darkening the skies of the densely populated East.

The best way to keep the soil from blowing away is to keep it covered all year around. There's a radical new idea for doing that: saving the land by bringing back the prairie—or, at least, by identifying and cultivating prairie-like plants which also produce food for us to eat.

The basic idea here is to try to find perennial plants which will continue to produce crops year after year, without the regular cycle of plowing and replanting that almost all our major food crops now require. Thus the soil is never bare to the winds; thus it stays where it can be farmed, instead of adding to the troubles of the dredge operators on our major rivers. Such perennials have been identified already. They do not produce yields as great as present grain crops, but the agronomists can surely improve them in that respect. They do require less cultivation and, above all, they do not leave the soil exposed to wind and water erosion between plantings. There *isn't* any time between plantings.

It is also possible to grow many useful food crops on trees. "Agroforestry" offers many advantages over field crops of all kinds, not least the presence of the trees themselves with all their beneficial effects on water retention, air quality and simple good looks. Particularly in undeveloped countries, it would be a good idea for many farmers to plant oak trees instead of corn: acorns can be made into a high-quality edible starch, and require much less effort to grow and harvest. And, of course, the trees do not need to be replanted in an average farmer's lifetime. Additionally, other kinds of food crops can be grown within the groves.

Trees can thrive with little or no irrigation in regions with low levels of natural rain. Their roots, which have been measured as far as two hundred feet below the surface, tap underground water sources that may be difficult to extract for irrigation . . . and (remember?) the carbon in a living tree stays there, out of the carbon-dioxide cycle, for decades and even centuries.

New Crops

There are literally thousands of species of plants (and animals) which can be used for human food, and yet barely a score of them comprise the great bulk of what the world's farmers produce. Some of the neglected species, like the winged bean, are among the few plants grown by man that are totally edible, leaves, stalk and root. Others, like many African fruits, tolerate low rainfall without irrigation and yet are almost unknown in the rest of the world.

There are many unexploited edible animals, too. Unfortunately for American tastes, which don't like eating anything that isn't big enough to shoot or to catch with a hook and line, most of them happen to be insects.

The locusts which destroy crops are themselves a first-rate source of protein. So are many other insects. If we can persuade Mediterranean fruit-flies to be enticed into traps, instead of spraying them with malathion, they could be converted into, at least, an animal food supplement.

A good many of those neglected plant and animal food species may become a lot more important in the coming decades than they are now if the global warmup continues.

Farming in the Warmup

Agriculture, like everything else, will be affected by a continuing greenhouse effect. We have already seen that the corn and other belts are likely to move northward (in the Northern Hemisphere), but there are other probable effects as well.

As the atmosphere warms, there will be more evaporation of water into the air, which will inevitably be balanced by more precipitation. But that extra rainfall will not necessarily come where farmers need it, and some scenarios suggest that it will be even less likely to come *when* it is needed.

One strategy American farmers may employ to cope with drier conditions in the fairly near future might be to start growing dryland crops. A lot of work is already being done, particularly in Israel, in experimenting with dryland fruits and nuts that grow in the wild in Africa and elsewhere, and are eaten by locals but not cultivated.

Much research needs to be done in finding strains of these plants which can be grown successfully in other dry areas— temperature, soil and other conditions play a large part in their adaptability. But they can be found. And some may be very valuable even before the warmup. They could return productivity to such barren areas as the Kalahari Desert . . . and they could reduce the demand for irrigation in much of our own Southwest.

Even apart from its warming effect, the extra carbon dioxide in the air itself will affect productivity in two ways. First, extra carbon dioxide increases the efficiency of plants and actually *reduces* their water consumption. (But increasing warmth brings more transpiration of water out of the plants and back into the atmosphere, which will neutralize part of that gain in water efficiency, and may even mean that more water will be required. . . . Yes, it all gets very complicated, doesn't it? But it's still likely that most of the uncertainties will, at least to some extent, cancel each other out, so the general picture will be about the same.)

Second, the more carbon dioxide there is in the atmosphere, the more carbon plants have available to assimilate and metabolize into their own substance. Thus the faster they will grow.

That doesn't mean that all plants will experience extra growth across the board. It will vary from location to location, since growth will be affected even more strongly by the accompanying changes in climate, many of which will neutralize any gains from extra carbon dioxide. Still, most plants in the world will grow taller and faster.

But carbon, like nitrogen, is only a part of what food crops need to supply us with nutrition, and as we have seen in the case of fertilizers this may merely mean that the plants will grow bigger, without necessarily providing more nourishment.

Finally, there is one sure way we can get more of the food we need from our farms, and that is by taking better care of it after it is harvested.

As much as a third of all the food grown in the world is spoiled by pest depredations and by poor distribution, so that it is harvested in good condition but rots before it reaches the consumer.

We certainly can do a lot better in all of these respects—and preserve the environment in the process.

18

How Hard Will It Be?

Now that we've seen what we have to do, we turn to a hard question. How do we do it? How do we persuade the majority of the world's people to change their ways? How do we get laws passed to make the changes to do the job?

All that seems very difficult, so perhaps this is a good time for us to cheer ourselves up with a look at the bright side. Yes, Virginia, there *is* a bright side. After all the time we've spent in looking at the way things have gone wrong, it's worth reminding ourselves that there are a few places, here and there, where they've gone very *right*.

So let's reward ourselves with a look at one of those very right places.

The name of the place is Kalundborg, and it is in Denmark. Kalundborg is a small city which contains a number of industrial plants, all of them of the kinds that we have identified as serious polluters and contributors to our environmental problems. There's the Asnaes power plant, the biggest electrical generating installation in Denmark. There's the Gyproc factory, which produces great quantities of plasterboard for construction. There's an oil refinery, and the Novo Nordisk pharmaceutical plant, as well as the usual government and retail establishments, homes and a surrounding collar of farms.

All of them have their special needs for raw materials and other resources, and all of them produce their own wastes. Novo Nordisk needs process steam to make its enzymes and drugs. Until about ten years ago, like everybody else in the business, it made its own by burning fossil fuel and boiling the water it took from the local rivers. Then it made a deal with the Asnaes power plant to buy their waste steam—still very hot, but no longer sufficiently super-hot to turn their turbines—which, like most power plants, they had previously simply allowed to cool and discharge into those same rivers, to their detriment and the detriment of their fish.

Asnaes had been drawing its steam water from those rivers, but then it turned to the Statoil refinery (which had some warm water left over from cooling its oil-processing towers) and bought that water from them. Statoil had something else to sell, too. There are waste gases from oil refining, usually not enough of them to be economically worth collecting and selling, and generally they are flared off in permanently burning torches. Statoil sold those gases—first to Gyproc, to replace other fuel in making their plasterboard, then to the power plant to buy up the remainder.

The Asnaes power plant still had surplus heat, so it began piping it to the town's domestic-heating system; this permitted some 1500 homes in Kalundborg to close down their household furnaces. There was still more heat at the power plant, in the form of relatively low-temperature hot water, so they used that to supply a nearby fish farm.

In fermenting its pharmaceutical products, Novo Nordisk produces a lot of thick sludge. They don't discharge it into the rivers, as many such factories do elsewhere in the world. Instead, they treat it to kill off microorganisms and then sell it to the local farmers as fertilizer. The farmers love it. They can use it without fear of heavy metals or industrial toxins; there never were any there. Fly-ash from the power plant makes cinder blocks, and the plant's new scrubbers will provide cheap gypsum for the Gyproc plant to make into plasterboard.

And it wasn't even primarily concern for the environment that started Kalundborg on its green path. Each of those companies was looking for ways to increase their profits . . . but Gaia is pleased.

Kalundborg hasn't solved all of our environmental problems. It hasn't even solved all of its own. But it shows how many of them *can* be solved in ways that actually make money and keep people employed, as well as doing essential work in the repair of Planet Earth.

That's the bright side. It's not all bright.

Make no mistake about it, our environmental problems mean that large-scale changes lie ahead. Businesses will be harmed, people will have to change their jobs. The reason for this isn't that do-gooder environmentalists like ourselves insist on it because of some idealistic devotion to "nature" or the spotted owl. It's because our profligate ways have done so much harm that large-scale change is *inevitable*. The only choice we have—the only future we can invent—lies in deciding which *kinds* of change will be best in the long run, the ones that will come about because we try to clean the world up, or the worse ones that will come about on their own if we don't.

The fact that many people will lose their jobs is bad news for them. It isn't good news for anybody else, either.

Unemployment anywhere is a drain on the country's resources everywhere. Adding to it is not a plus. This growth in
joblessness won't happen because anyone wishes it to. It will
happen inevitably, simply because there is no way to avoid it.
If we drive our cars less, they will wear out more slowly and
fewer cars will have to be built to replace them; therefore
jobs will be lost in Detroit (and in Osaka and many other places
around the world). If we recycle paper, fewer trees will have to
be cut down to make new pulp; whereupon many of the men
and women whose jobs depend on lumbering will lose them.
If we cut down on the burning of fossil fuel, oil workers and
coal miners will be laid off.

But if we *don't* do those things, we face a future of disease,
scarcity and discomfort . . . at best.

Those of us who are not directly personally affected by the
changes environmental dislocations will bring about can
take some philosophical comfort from reflecting that all these
things are going to be happening in a good cause. That isn't
likely to cheer a newly unemployed person up, but there's a
bright side here, too. Although many jobs will disappear,
many new ones will be created and more often than not the
new jobs will be better than the old.

Does the local automobile factory close down because no
one's buying new cars right now? Too bad; but the fact that
people don't want to drive cars very much any more doesn't
mean they're willing to quit traveling entirely. They'll be
customers for the trains, perhaps the magnetic levitation
trains we talked about earlier. Somebody is going to have to
build those maglev trains, as well as the light street rail systems and the monorails and the new fuel-stingy aircraft. If
the car company's managers deserve their opulent salaries,

they probably will have diversified in those directions before the car market dried up entirely. After all, they did so very well once before, when under the stresses of World War II they switched over to a completely new product line of tanks and Army trucks and bombers as easily and successfully as they had made cars before. The major manufacturers who convert in time will flourish. The ones who don't won't.

Then there are all those homes to retrofit and the new ones to build. There's all that national infrastructure—the decaying water and sewage systems, the rusted and worn bridges, tunnels and public buildings—which need to be re-built before they collapse entirely from their present decades of neglect. If lumbering slows down to a crawl, there are whole huge new industries to create in the Pacific North-west, like fish-farming in the mouths of the great rivers, or building and tending power-generating windmill farms, or even "agriforestry" to provide food from clear-cut lands. There is a great need, which will surely become a greater one as our population ages, for health workers of every kind, from paramedics to RNs or even MDs.

Some of the myriad new jobs will not attract the skilled worker, but we have all those *un*skilled workers who are now cut out of the job market entirely. For them there will be service jobs, some traditional, like working in fast-food stores, some relatively new, like sorting trash for recycling. The so-cial value of creating jobs of this sort is immense; it can convert welfare clients into productive wage-earners. But probably most of the new jobs will actually be better ones than the old, at least in the sense that they are less damaging to the health and the spirit than mining coal or working on a heavy-industry assembly line.

THE TECHNOCURES

So there will be plenty of new jobs.

We should face the fact, though, that such consolations will be of little comfort to the man or woman who is put out of work in the prime of life, since he or she will have no guarantee of getting one of those better jobs—if any job at all. The only certainty for these unemployed is that their painfully acquired working skills and experience will never again be wanted.

The other fact we must face is that, sadly, there is very little chance that any of us will see any real benefit from all our environmental efforts right away.

It is impossible for us to make things instantly *better*, no matter what we do. We can't. The best we can achieve is to prevent them from getting unspeakably *worse*.

The damage to the ozone layer won't go away; in fact, it will increase, at least for a time, no matter how hard we work to limit it. Nor will the global warming stop, nor our destroyed soil and water return at once.

Even so simple, and local, a thing as cleaning up a body of water won't produce immediate results that we can see. If you somehow manage to make your nearest lake sparkly clean, it won't stay that way. It will soon be contaminated again by airborne pollution unless a great many other people, living hundreds or thousands of miles away, also act.

Indeed, in the case of some very highly polluted rivers, the first effects of a cleanup may make them look worse than ever before. As the cleaner waters begin to dissolve out the accumulated sediments of generations at the bottom of the riverbed, lumps and clumps of filthy pollution are likely to break free and float away on the surface.

And the final, in some ways the scariest, consequence of

environmental action we must face is the ripple effect from the inevitable economic changes.

Let's take a deep breath, brace ourselves and do that now.

Let's suppose that tomorrow the American government decides to put the necessary environmental measures into effect at once. Congress passes the necessary bills, and the President signs them into law.

Let's say that in that first legislative bundle are such relatively moderate, preliminary measures as a total ban on all CFCs, a fifty-cent-per-gallon tax on gasoline to cut consumption, a stiff requirement that the cars of the near future average 50 miles per gallon, a three-cent tax on every kilowatt-hour of electricity produced by the burning of fossil fuel and even a cutback on military excesses by abandoning a dozen new weapons systems, phasing out many bombers, missiles, tanks and warships, reducing troop levels and closing many military bases.

For starters, can you imagine what that kind of simultaneous bad news about autos, oil companies, public utilities and defense manufacturers—the bluest of blue chips—will do to the stock market? The October, 1987, one-day drop of more than 500 points in the Dow-Jones averages—which meant that one-fifth of the dollar value of every American stock investment disappeared in a single day—would look like a minor "technical correction" by comparison. Even the Great Crash of October, 1929, might pale alongside that sudden fall.

Can you imagine what the loss of all those jobs—many of them very high-paying jobs, too—would do to the national unemployment figures?

As long as we're making our blood run cold with worst-case scenarios, try imagining, too, the consequences that will inevitably follow as some of these suddenly unemployed homeowners can no longer meet their mortgage payments and are foreclosed; as upper middle-class families, accustomed to having considerable disposable incomes but now jobless, stop buying new cars, TV sets, VCRs, household appliances, clothing and furniture—cut back on travel and vacations—begin to sell off their own accumulated savings, in the form of stocks, bonds and mutual funds, so that they can meet their living expenses . . . and thus further contribute to the plunging market?

Can you imagine a Congress voting for, and a President signing, laws that will do all that to the economy?

And even if we can stretch our imaginations that far, so that we can persuade ourselves that the United States would be willing to take the first dose of this unpleasant medicine, what about the rest of the world?

For that may be the hardest part of all. If it seems that it will be difficult for America to change, consider what those changes mean for most of the other 140-odd countries in the world.

The environmental problem is not an equal-opportunity threat.

For example, the effects of global warmup—even severe warmup accompanied with the whole panoply of ills like high sea-level rise and violent weather—aren't going to affect everybody in the world to the same degree. Florida may have to worry about how high the ocean tides will rise. Switzerland has no such fear. If the American midwest suffers crop losses from a radical climate change, the Sahel, for instance, may actually find that in the changing precipitation patterns their own climate has improved.

In that event, what should we expect? Can we ask the Sahelians to join with us in preventing what will actually for them be a blessing? And if we do ask, will they agree? Will they, for instance, display enough charity to overlook the fact that we, not they, are the ones who have actually created the problem?

To make a difficult problem even more complicated, it is unfortunately true that some parts of the human race may find themselves better off with a limited amount of global warming. It is also true that, as in the case of the Sahel, a good many of those areas are presently impoverished and indeed desperate in the face of the living conditions they now endure. How do we persuade those people to cooperate? It is a good deal to ask of any nation that it turn down a possible improvement in its own miserable lot for the sake of keeping our comparatively fortunate one from getting worse.

Even for Third World countries which would not benefit from the climate changes, we can't realistically expect them to stay undeveloped simply because it suits our convenience.

Some moralists might raise that as a question in ethics: Do we Americans have any right to ask an African, Asian or Latin American peasant to forswear any hope of ever owning a car or a dishwasher, or of flying in a jet plane? But as a practical matter, whether moral considerations would let us do that or not, we won't have the privilege of making that decision.

Whether or not we ask them to abstain, few Third World people are going to be willing to abandon their hopes of making their own lives better for the prospect of some abstract future good—not while they can see every day, on their little black-and-white village television set, that the rest of us are enjoying copious quantities of these things already.

It is not merely peasants who will be less than overjoyed at the prospect of making economic sacrifices. Some highly industrialized societies are in similar straits, if not worse ones.

The most conspicuous of them are the nations of Eastern Europe. Miraculously and wholly unexpectedly, they threw off the state Communism that ruled them for nearly half a century. It wasn't entirely ideology that caused that astonishing revolution. The desire for free speech, free elections and all the other freedoms certainly played a part. But most of all the driving force was simple hunger for a better life—for better food, more conveniences, more of all worldly goods—for, in short, the good life of material well-being that Communism had promised and so abjectly failed to provide.

We have already seen that the industries of the countries of the former Warsaw Pact are notoriously the dirtiest and most destructive in the world, and the ones most in need of reform for all our sakes.

And yet, if it comes to a choice between either cleaning their industries up or, alternatively, producing more goods for the people of those countries—as it surely will at least in the short term—which way do you think they will decide? For that matter, which way would *you* decide?

What we have here, in effect, is one more example of the unhappy economic principle which Garrett Hardin calls "the tragedy of the commons."

Hardin expresses it in an allegory, which goes like this: In a certain village twenty families live, and they share a common greensward which can pasture their household milk cows. The commons is just big enough to support exactly twenty cows. As long as each family puts just one cow out to pasture there, there will be plenty of grass for all of them, and all will have milk.

That's a good steady-state arrangement, as long as it stays that way.

If, however, one family puts *two* cows there, none of the cows will get quite enough to eat. The difference won't be large. None of the animals will starve, but their straitened diet means that each one of them will produce a little less milk each day. The total volume of milk the village's cows produce altogether will be pretty much the same, but it will be divided by twenty-one instead of by twenty.

What's the effect of this change?

It's an immoral one: selfishness triumphs. The selfish family that has put an extra cow on the common now has nearly twice as much milk as it had before . . . but each of the other families has to get along with a little less.

So, seeing this, enlightened self-interest causes each of the other families to put in a second cow as well. . . .

And the ultimate effect is that now there are forty cows on a pasture that can't support more than twenty. It isn't just a matter of lowered milk yields now. After a while the whole herd starves to death.

Each family, you see, has acted rationally according to its own best understanding of what will serve its interests. The tragedy that results is that the effect of all this "rational" individual behavior has collectively ruined them all.

In just the same way, a cooperative strategy for dealing with the assaults on our global environment will promote the general good . . . but it may be to the advantage of some nations and some individuals to act contrary to it.

That's the cheating problem in a nutshell.

Country X will well understand that it, along with all the rest of the world, is threatened by increasing carbon dioxide emissions . . . but its leaders may reason that if everybody *else* does the no doubt difficult and expensive things necessary to deal with the problem, the relatively small damage that will be done to the environment by the Xians won't make any real difference. Therefore X can coast along in the good, old-fashioned, high-polluting way—and be able to out-compete the rest of the world in the price of their export manufactures while they do it, since they won't have to pay the bill for the sacrifices.

How can we deal with that?

We can start by trying to persuade every country in the world to sign appropriate treaties, of course. But then what do we do if some countries cheat, or refuse to sign in the first place? Do we declare war on them?

The longer we look at the problems, the harder they seem. The only thing that spurs us on to try to solve them is that we don't have a choice: the costs of not solving the problems are even higher.

Still, we have to ask ourselves whether all these measures can ever be put into effect at all. Can we imagine the Congress of the United States—or any congress anywhere—taking the necessary legislative steps when they are sure to create that much economic turmoil?

To decide what's possible, we must look at what's been called "the art of the possible"—namely, politics; and that's what we will be discussing over the next few chapters.

THE
WAY TO GO

19

How Far Do You Want to Go?

We can start with the "non-partisan" kind of politics, because that's the kind we do-gooders are least uncomfortable with. (It's not the kind that can do the most good, though, but we'll talk about that later.)

All the same, some sort of action is the next logical step for us, because now we've come to decision time.

Since you've got this far in the book, probably it's reasonable to assume that you agree that at least some of these Ghosts of Times to Come have to be taken seriously.

The next question is, what do you want to do about it? How far are you willing to go beyond the usual daily activities of your life for the sake of trying to preserve a decent world for your grandchildren?

The sorts of commitments that you might choose to make come in all shapes and sizes. You can start off with some fairly small ones; there's a lot you can do in your personal life that is worth doing.

The general watchword to describe your personal contributions to a better world is "restraint": Use your furnace and air conditioner—and your car—more sparingly, when possible substituting a sweater, a fan or public transportation instead.

Turn off lights when you don't need them. Help with the ozone layer by avoiding (for instance) foam-plastic cups. Cut down on waste by recycling everything recyclable—or by reusing some of the things you're expected to throw away: if your supermarket bags your groceries in plastic, you can save the bags and bring them back to be used again next time you shop—the bags don't weigh much, and they last a long, long time. (Or you can buy a couple of string bags like the ones your grandmother's mother used to carry.)

These are all good things. You can justifiably feel pleased with yourself if you are doing them . . . but it does not lie within your power to do everything that has to be done by your solitary self. As we've also pointed out before, some of what has to be done has to be done on a far larger scale than the individual.

If we want to *solve* the problems, we're going to need a lot of help. It will have to come from among others, the politicians who run our governments, because the job can't be done without the passage, and enforcement, of some tough new laws.

This has a thoroughly nasty sound. You don't have to be a Libertarian to object to the idea of Big Brother standing over us to make us do what's good for us. So before we commit ourselves to dirtying our hands in the political mudbath, let's double-check to see if there are any possible alternative choices.

Actually a number of alternatives, of varying efficiency, do exist. For instance, one of the best nonpolitical ways that you might employ to try to persuade industries to mend their ways would be to hit them in the pocketbook. If enough people refuse to buy polluting products their manufacturers will have to clean up their act or, sooner or later, go out of business.

Trying to employ "market forces" to effect environmental change has had some success in Europe. In England the "Buy Green" campaign has brought about the sudden appearance of unbleached paper products and ozone-friendly appliances on supermarket shelves. In Germany, the Greens have even set up a sort of *Good Housekeeping* seal of environmental approval, awarding the right to put their emblem on the labels of the most benign products and withholding it from the polluters.

The United States, though, is running a long way behind in this respect. Although attempts are now under way to establish some sort of environmental seal for merchandise, we're far behind some of the Europeans and we certainly have no powerful Green Party to vote for at the polls. So, even if we want to reward the environmentally conscientious and punish the evildoers, how can we? If we want to shop for a new VCR, is there any practical way for us to know whether its manufacturer has rinsed his circuit boards with ozone-killing CFCs or with harmless terpenes? Can we tell whether the brand of white paper napkins we see on the supermarket shelves were bleached with dioxin-producing chlorine, or by the fairly harmless oxygen process?

We may think that, at least some of the time, we can tell from the labels. The reason we are led to think that is that a lot of manufacturers are way ahead of us in recognizing the cash value of the environmentalist consumer.

They're not all doing anything to *deserve* our loyalty, though. Almost all the labels are thick with weasel words.

You will often see a spray-can marked "no chlorofluoro-carbons" (but without saying what other ozone-killer may have replaced them) or find a paper or plastic product boldly labeled "fully recyclable" (which claim may well be true about the *possibility* of recycling, but tells you nothing at all about whether any system is in place to actually *do* it).

Sometimes such "dealer's talk" (as advertisers call the statements they know aren't true, but hope nobody will make a fuss about) may be trustworthy. It also may not. We know from sad history that even presumably respectable major corporations have been known in the past to do such things as sell colored sugar water as orange juice for babies, put addictive narcotics in soft drinks and market pablums as disease cures. Most chicanery of the sort that affects health is now illegal. Sometimes (though not very often) it is even punished; but there are no serious laws to prohibit or punish misleading claims about environmental qualities.

As a last resort, then, we might try to take the policing of our suppliers into our own hands.

If we were determined enough, conceivably we could carry test kits around with us for everything we eat or drink or use, and we could analyze the emissions from each factory, and test for toxins the effluents they discharge into every body of water. . . .

But, really, if we tried to do all that, would we have any time left to *live*?

Some environmentalists have gone a good deal farther in taking the preservation of the environment into their own hands—sometimes violently. Canada's Sea Shepherd Conservation Society rammed and damaged an illegal whaling ship off the coast of Portugal and scuttled two other whaling vessels in the harbor of Reykjavik, Iceland; another Canadian group blew up a hydroelectric installation on the island of Vancouver. In Thailand a group set fire to a chemical plant; in Australia activists destroyed more than a million dollars' worth of lumbering machinery; in Germany their specialty is knocking down high-voltage transmission towers.

Perhaps the most famous of the radical environmentalist groups, though, is America's own Earth First! Earth First! (the exclamation point is an integral part of the name) has been described as an organization of ecoterrorists, and at least some of its "leaders" (Earth First! has no elections and indeed keeps no membership rosters) are quite extreme. For some, environmental action isn't for the purpose of protecting human beings, it's to protect the world *against* the human race. "All species are equal and have an equal right to exist," according to one statement of its philosophy; another leader proclaimed that such disasters as famines and the AIDS epidemic were valid measures for getting rid of "this human pox that's ravaging this precious planet."

Most of Earth First!'s activities are conducted with one eye on the television cameras, and some are pure eco-theater—even eco-comedy, sometimes. (As when, on March 21st, 1981, a group of seventy-five of them drove to Glen Canyon Dam in Arizona. While most of them staged a noisy demonstration before an audience of tourists and Park Service police, a few scrambled to the top of the dam and unrolled a plastic sheet with a long, black crack painted on it; when the tourists caught sight of it they concluded the dam was about to break and there was panic.)

Some of Earth First!'s actions are along the same general lines of other environmental groups, like Greenpeace: intervening with whaling ships, confronting polluters and so on. Some are a good deal more violent. Earth First! members claim to have, or are accused of having, blown up bulldozers, driven six-inch spikes into trees scheduled for lumbering, pulled the survey stakes up from planned logging roads and poured grit into the lubrication systems of road-building machines. These acts of ecological sabotage—they call it "ecotage" or "monkeywrenching"—have stopped short of killing anyone, so far, but they may have come close. In May, 1987, one spiked tree, going through a lumber mill, shattered the blade of a band saw and nearly cut a mill worker in half (but Earth First! denies that it had spiked that particular tree). One Earth First! leader, lying on the track before a locomotive, lost both legs when the engine didn't stop. And on May 24th, 1990, two Earth First!ers, Judi Barr and Darryl Cherney, were blown up when a pipe bomb exploded under the seat of their car. Authorities claim it was their bomb, accidentally set off; Earth First! claims that it was planted by anti-conservation interests, probably lumbermen. Both survived, though Cherney suffered internal injuries and a broken pelvis.

It is easy to recognize the frustration many environmentalists feel when they see heedless destruction going on with impunity; but it is hard to believe that the answer is more destruction. When we're trying to achieve moral objectives it seems better to use moral means—even if they're slow, arduous and uncertain. And that means going through lawful channels.

* * *

So when we come right down to it, here too we don't have any choice.

Saving the world is not a cottage industry. To get the job done we have to have laws and enforcers. We need government to step in.

That means getting into politics—one kind of political activity or another.

Which brings us back to the question of just how far you are willing to commit your life, your fortune and your precious honor to the cause of making the world safe for your grandchildren to grow up in.

You have a choice of a wide spectrum of activities. You can choose to be a missionary to the unsaved, or an organizer, or even a political boss.

All these activities are well within the powers of any average adult human being; and in the next three chapters we'll show what they can accomplish, and we'll even give you step-by-step instructions on how you yourself can go about doing them.

20

Missionary Work

All right, let's suppose that by this time you have gone about as far as you can go on your own.

You've segregated your trash for recycling, you've oiled up your bike for short trips to the drugstore, you've made a mulch pile for kitchen waste and grass clippings and you've turned back your thermostat. It doesn't feel like quite enough. You still want to do more.

What's the next thing you can do?

You can do a lot. You can start trying to get other people to follow your good example.

That sounds as though it might be hard work. It isn't really. You can do a lot of it in your spare time, without even leaving your home; all you need for the purpose is some writing paper and a few postage stamps:

You can write letters.

That may not strike you as the kind of thing you would normally want to do. We all have a picture in our minds of frustrated old codgers with nothing else left to do with their lives, firing off crank Letters to the Editor or to a random celebrity whose address they can find. You're sure that that's not *you*. You certainly don't want other people to get the idea that you're some kind of a *nut*.

Please be reassured about that. You aren't a nut. On the contrary, trying to educate the world to the dangers of what we're doing to the environment is being unusually *sane*. The preservation of our planet is a matter of universal concern, and it's *important*.

You don't even have to sound even the least bit nutty when you write your letters. You can do it quite easily and with your reputation intact. If anything, your reputation will be better than ever, because from what you write it will be apparent that you're a good citizen who is only doing what every good citizen damn well *ought* to do when this kind of catastrophic threat looms before us.

Not only that, but many of the people you write actually will *want* to hear from you—because they have a big interest in keeping you happy, and your letters will tell them how to go about it.

For instance, a good choice of people to write to are the people who make and sell the things you want to buy.

Let's say you have a favorite brand of toilet paper, sanitary napkins or paper towels. If they're white or tinted—if they're any color but brown-bag khaki—they're probably made of chlorine-bleached paper. We know that's bad stuff. You really don't want to encourage it, so there is every reason in the world for you to write a little note to the manufacturer, informing him that you would like to go on buying his product, but that you'll have to switch to another brand if he doesn't stop pouring dioxins into the environment. A letter might go like this:

Dear Sir:

I have been a customer of your product for years and would like to continue buying it. However, I don't want to go on contributing to environmental damage. Isn't it possible for you to switch to unbleached paper, and preferably recycled paper, in the future? Do you have any plans for doing so?

(It wouldn't hurt to enclose a label from your most recent purchase of his product, to show that you should be taken seriously.)

Or if you're planning the purchase of some electronic gadget, you can write to several manufacturers:

Dear Sir:

I intend to buy a new computer in the near future, but I don't want to buy anything made by the use of CFCs or other ozone-destroying chemicals. Please tell me if this is the case in your products.

Daily newspapers are major contributors to paper waste and to deforestation—a single Sunday edition of a large metropolitan paper costs the world 50,000 or more trees cut down to make newspulp. Write your own newspaper a little note:

Dear Sir:

I'm a subscriber to the *Herald*, but I am considering cancelling all subscriptions to papers which do not make use of recycled paper instead of contributing to the destruction of our forests. Can you tell me, please, what your plans are in this respect?

As you see, the tone of your letters can be firm without being nasty; and the letters you get back will surely be equally polite. Your single letter may even help persuade your correspondents to start in the direction of making real changes. If they get a *lot* of such letters, have no doubt about it, they definitely *will* make the changes.

The same sort of letter can go to every manufacturer and even every retailer whose customer you are—or might be—protesting excess packaging and environmentally destructive manufacturing processes. Whatever they do that you don't like, *tell* them about it. Some of them would be happy enough to change, given some incentive; they simply have not seen any evidence that anyone cares.

Then there's that long list of pen-pals you already have: the relatives, the old school chums, the former neighbors. You're going to write them a letter every now and then anyway. While you're at it why not let them know how you feel about the damage to the environment?

You can do it easily enough, even off-handedly. Suppose there's a story in the paper about the garbage crunch, or pollution in your local water, or recycling—as there surely will be, in almost every newspaper in America, almost every day. Make a photocopy of it and stick it in your letter, and add a paragraph:

> Well, we're trying to do our part in recycling (or whatever), but as you see from the enclosure there's a long way to go. I worry about what kind of a world our grandchildren are going to live in.

If you have the habit of writing a lot of what's-been-happening-at-our-house letters at Christmas, you can easily include something with them. (For instance, you might draw up a list of ecological New Year's Resolutions for your family, duplicate it and send a copy with each letter.) This sort of enclosure is what mail-order companies call an "envelope stuffer"—it costs almost nothing to send it out, since very few letters reach the extra-postage limit of an ounce, so you might as well stuff your envelopes with something that may do some good.

And then there's the time-honored practice of writing letters to the editors themselves.

To be sure, if you're a normally retiring person this is a considerable step beyond anything we've suggested before. It means coming out of the closet, exposing your private beliefs to the world at large.

Going public in this way isn't in any way *wrong*, though. You have an absolute right under the First Amendment to the Constitution to let people know exactly how you feel on public issues . . . and some people would even argue that it's your duty to do so, as well.

When should you write a letter to the editor? Whenever you want to, but most effectively when there has just been some relevant news story or remark in a column that catches your attention:

Dear Editor:

Your editorial opposing the bond issue for the new sewage treatment system is a bad mistake. The whole country is trying to clean up its wastes; why should we refuse to do our part?

You don't have to stop with your local paper. What other publications do you read? Does your church newsletter, local shopping paper, tenants' association or homeowners' society publication have something to say that you don't agree with? (Or that you do?) Then let them hear from you. And what about the national magazines? For instance:

> Editor, *Time:*
> In your coverage of the recent presidential order on cleaning up military toxic wastes, you didn't mention that some of those wastes are already polluting the drinking water for hundreds of thousands of people around the country. The order doesn't go far enough. We need to have Congress provide more Superfund money, and we need to make sure that the appropriated money isn't held back by the Administration but is actually spent to do the job *now.*

Perhaps most important of all, let the people who will *make* the important decisions for you know how you want them to make them.

These people are your congresspersons, senators, president, vice president, governor, state legislators, mayor, local or county officials, members of zoning boards, sanitary commissions, park departments or whoever else is in a position to cast a vote or make a decision that affects the environment.

There are a lot of these people. Almost all of them have to make up their minds almost every day on some such issue. Quite a lot of the time they won't have any strong, preconceived ideas of which way the decision should go.

The fact that they're not sure what to do doesn't get them

off the hook. When push comes to shove they'll have to make the decision one way or another anyway. Your letter might just tip the balance. Some of the rest of the time they may have their minds made up firmly enough in advance—but if they get enough of a groundswell of pressure in the other direction from the public they may well change their minds after all.

What should you write them about?

If there is a specific issue you know to be coming up before them, then get specific:

Dear Zoning Board Member:

I urge you to vote no on the application to erect 52 new one-family houses on the old Smith tract. We need more open green space and woods, not more construction.

Or:

Dear Congressman Brown:

Please support HR XXXX, calling for the preservation of the Alaskan National Forests. We can't criticize Brazilian peasants for cutting down their rain forests when we are doing the same thing with one of the few remaining virgin stands of Northern hardwoods in the world.

How do you know who to write to? Newspaper stories will tell you what governmental agencies are involved in any decisions; if you then don't know the names of the people who serve on them, and want to address one or more of them by name, call your city, county or municipal clerk's office and they will tell you. (You can find the clerks' numbers in the phone book.) Presumably you already know the names of your two senators, your congressperson and the president of the United States, or at least you know enough to look for them in the daily newspaper.

Where do you send the letters? To the President, the White House, Washington DC will get there. To senators, "Senator James White, Senate Office Building, Washington DC" will make it; to congresspeople, "Representative Bill Brown, House of Representatives Office Building, Washington DC" will do the trick, too. It's better if you know the zip codes, which you can get by calling their local offices (all representatives and senators maintain offices in their home areas) or your local post office. But it's not essential. For other officials, the same county (or other) clerk who gives you the names will also give you the proper addresses on request. Don't hesitate to make the phone calls out of some kindly wish to avoid bothering them. That's what the people who work in those offices are paid for—by you!—namely, to serve the people of their district—like you.

It may occur to you that your single letter to a senator won't be taken seriously.

Surprisingly, that's quite wrong. The senator is, of course, not all that likely to read and ponder your letter himself. But he employs a crew of people on his staff for no other purpose than to log the incoming mail. At the least they will surely give him a tabulation of how many constituents have written him on which side of all the current controversies. In lower offices, where the office-holders get a lot less mail, the people you write to very often *will* read your letter, and what's more pay attention to it. Your arguments may not sway them to your position . . . but if you don't let them know what you want at all you *know* you won't sway them.

On the other hand . . . it's certainly true that two letters carry twice as much weight as one, and a few hundred letters carry a *lot* of weight with almost any elected official, even a president. Perhaps you would like to multiply your forces by enrolling others in your activities.

It's easier than you think. The next chapter tells you how.

21

Leverage

By the time you've used up your first roll of stamps you have reason to feel pretty proud of yourself, but maybe you feel a little lonesome, too. You can accomplish a great deal as a single voice in the wilderness. All the same it would be nice, you think, to have more company.

So let's take a lesson from Wall Street.

When the smart financial sharks set out to take over a corporation, they don't use their own money. They use other people's. They borrow it from individual investors, banks or savings and loans, or they get it by issuing the famous "junk bonds." The only thing they do with their own money is use it to promote, and then to control, the investment capital borrowed from others.

That's how they get far more clout for the buck, and the name of this process is "leveraging."

The potentialities of leveraging are not limited to the amplification of money. You can leverage your efforts on behalf of the environment, too.

The way to do that is what activists of all kinds have been doing since time immemorial: to join (or even to start) an environmental organization in your own area.

It's not that hard to do, and it has big advantages. There's the social advantage: you will meet new people whom you will probably enjoy knowing, since they will share a lot of your feelings. There's the leverage advantage: if the organization never gets larger than half a dozen people, that still means there will be six times as many people doing what you have been doing on your own, and thus six times as effectively.

And never forget that the biggest advantage of all is that you just might save the Earth.

Joining Up

Before you start organizing a pro-environment group in your area, check around. You may be able to save yourself some effort. There may be one already. There may even be several. If there are and you have never heard of them, that means they aren't doing a good job and probably need your help.

One way to find out if they exist is to write the various national or worldwide pro-environment groups (Friends of the Earth, Greenpeace, etc.—there's a list of them at the end of this book) and ask them if there's any local group in your neighborhood. Another way is to pick up the telephone and ask someone at your local newspaper. If the paper is reasonably large it will likely have assigned someone on the staff the duty of keeping track of such organizations. If it isn't, almost any reporter will probably know.

Let's say, though, there's nothing. How do you go about starting something from scratch?

Mailing List

You locate a few people who seem to share your views and you send them a letter inviting them to help you organize a club.

How do you find these people?

You probably already know some of them, especially if you've been speaking up for a while in letters to the editor, or in church and other meetings. For new names, you watch the letter column in the paper, looking for letters from people whose views seem to match your own. When you find them, you look up their addresses in the phone book. (You should include a zip code, and that won't be in the phone listing. The letters will probably get there anyway, but the mailman will appreciate the zip—appreciate it enough so that if you take the list to your local post office they'll tell you what the appropriate codes are.)

All of these names constitute your mailing list. When there are a dozen or so names on the list, you send them all the form letter:

Dear Friend:

If you're as concerned about our environment as I think you are, let's talk about what we can do about it. I'm inviting a few others to join me in my home for coffee and a discussion at eight o'clock on Friday, the 9th of April, to explore the possibility of starting a local environmental group.

Please come if you can—and by all means bring a friend if you like.

That's a minimalist letter. You can embroider it any way you like.

You can make it better if you put the actual name in the salutation instead of "Dear Friend"—that's an old mail-order trick. (Not hard to do, either, if you happen to have a word processor to do your typing for you, with a "mailmerge" program for form letters.)

You can make it more urgent and timely if there has been some recent event in your area that has made an environmental impact (an oil spill, a proposal for a new garbage incinerator, a particularly noxious fire in a junkyard for car tires). If so, you can tie your letter to that with even more effect.

So you send out your letters and then, when you get all these people together ("all" perhaps meaning no more than three or four), you suggest you all turn yourselves into a club.

That's all there is to it.

Along about then, you may find out from one of your new contacts that there is such a club in your neighborhood after all; it just hasn't made enough impression on the world for you to have heard they exist. Then you can all decide to join that one—or at least to see if it has a meeting you can attend so you can check it out.

If that doesn't appeal, or if you give it a shot and decide the existing organization doesn't appeal to you, you've already got the nucleus of your own anyway, and you can start right in on Doing Things. The next chapter tells how.

22

Organization and Action

Now you've got your environmental club. What are you going to do with it?

For convenience, we'll divide the club's activities into three main sets of things to do. The first thing the club should do is *hold meetings*. Try to do that once a month, and try to keep doing it *every* month.

The second thing the club does is *grow*. Keep reaching out for new members; try to find at least a dozen new names of prospects to invite for each meeting.

The third thing the club does is *do things*—the sorts of things you organized it for in the first place. Start to have an effect on what goes on in your area.

We've put this "action" item third on the list, because it seems a more logical way of organizing what we have to say on this subject, but it shouldn't be third in your priorities. The best way to have your organization survive and grow is for it to *act*. You should start some sort of action as soon as physically possible—at your very first meeting, if you can, even if it's nothing more than having each person present sit down and, there and then, write a letter to the President of the United States.

Let's take this three-part program up one part at a time, starting with exactly what it is that you do at the meetings.

Meetings

The first requirement for a meeting is a place to have it in.

You'll probably start out in somebody's family room, but unless you're very unlucky that should soon become inconvenient. Then you need more space. That won't be hard to find; go to your local church, school, library, firehouse or community center and ask them about borrowing (or renting) a room.

What you need for a meeting room is enough space to hold everyone who shows up (but not so much space that they rattle around—there are few things more deadening to a meeting than a lot of empty seats), a location that isn't too hard to find and the privilege, if at all possible, of serving coffee and so on to those present. You may want audiovisual equipment, or microphones, for some kinds of programs; check to see if they're available, or if you can borrow them and bring them in.

Program

Since you're going to have meetings (because without them you won't have a club at all for very long), it's a good idea to make them as interesting as you can.

Your best bet for achieving this is to present either a speaker or a film at every meeting.

For a speaker, see if you can find some reasonably articulate person to address a different topic each month. Topics like: Is your local public utility doing its best to avoid pollution? Is there a radon problem in your area's homes? How safe are your local water supplies? Is any local factory producing toxic wastes?

When you get big enough, you can attract real experts with name-recognition value; failing that, you can ask one of your own members to draw up a report. Even a book report can be interesting—somebody reads some environmental book, and tells the meeting what it says. Keep the talks short—twenty minutes is a good length of time, with as much added time as needed for questions, discussions and so on afterward.

A short film (or videotape—same thing) can substitute for a speaker. ("Short" means no more than half an hour—here too, twenty minutes would be better.) For help in locating useful tapes or films, ask your local librarian, public or college. Almost all libraries will have catalogues of countless films of all kinds on all subjects. In addition, the libraries themselves may well already own some useful films that they can make available to you.

The speaker speaks, the film rolls; when the feature is over you have a discussion on it. Then maybe some committee reports on activities of the group. You don't need much more than that. Ninety minutes of program all told is plenty. Then you serve coffee and cakes if you can, and let people chat and mingle until they want to go home.

Business

There are certain things you *don't* want to do at the meetings, or at least that you definitely don't want to do any more of than you can help. Such things are the deadly-dull (and frictional) organizational matters that plague most organizations.

In particular (though this may sound strange and antidemocratic) you really don't want to spend one minute more than is absolutely necessary on such things as writing a "constitution" for your club. The reason for that is that the writing of constitutions and by-laws is not only a terminally boring process for most people, it is a preposterously divisive one. You will be dismayed to find how rapidly your friendly little group splits into unfriendly little factions to argue over every line in the constitution . . . if you let it happen.

Do your best to get that chore out of the way as fast as you can. At the first meeting appoint someone to write it; at the second, pass it; thereafter, do your best to leave it alone. (If you don't have anyone to write one, a model constitution appears in Appendix 2.)

At some point you may want to incorporate the organization. This is a judgment call. The legal formality of incorporating is a little bit of bother, and it costs money, though not usually very much. But it regularizes your handling of whatever money the club may handle, and has the benefit of limiting possible liability lawsuits if anything should go seriously wrong at a meeting. If one of your members is (or knows) a benevolent lawyer, ask him for help.

You will probably want to elect some officers. Do it as quickly as you can, but preferably only for short terms at first—let the election of long-term officers wait at least for four or five meetings, so you can have a chance to see who is good at what (and so you'll have a larger selection to choose from).

Expansion

Now your club is in being. The next question is, how do you make it grow?

Basically you do that the same way you attracted members in the first place. You build your mailing list by keeping an eye on letter columns, etc., for prospects, and then sending them invitations to meetings; and by urging everyone present at any meeting to invite others, or at least to supply names to send invitations to.

At each meeting you might want to spend ten minutes on recruiting as part of the program. You take time out for the purpose, have on hand a supply of envelopes, stamps, writing paper and printed invitations, and urge each person present to send a little note to as many people as he can think of for the next meeting. Some people who are active in church or other groups can make announcements, or get announcements published in their newsletters.

But it's still true that the best way to grow is to start *doing* things. The more visible your club becomes through activities, the more potential members will become aware of your existence.

So let's look now at the third question: What *significant* things can the club actually *do*?

Saving the Environment

This is where you finally come to grips with what it's all about. All the organizational and programming stuff is a kind of foreplay. It's important. You have to do it, just as you have to mix the dough before you can bake the cake. But now you are getting ready to make things happen in the real world.

There are four main areas where you can undertake such purposeful group activity. They are not at all mutually exclusive; you'll do best when you do all of them at once:

Individual action. Encourage the members of the club to do the same things you've been doing as an individual, as outlined in Chapter 20. All that individual letter-writing and so on will be just as effective as ever, only now there will be more persons doing it.

Publicity. Whenever you do *anything*, let the world know about it. Send announcements of your meetings and activities to the local news media.

It's easy and cheap enough to do that. All you have to do is write up a press release and mail it out to your local newspapers, radio stations, TV stations (including the local cable community-service station if there is one), and any other outlet you think might be interested.

The form for a press release is simple, and goes more or less like this:

Bucks County Environmental Association
FOR IMMEDIATE RELEASE

Environmentalists Meet
At Harbinger School

The Bucks County Environmental Association will meet at the Harbinger School, Fulton Road and West Fifth Avenue, at 8 PM on Friday, May 21st. James Smith, author of the best-selling book, *Why Sunburn Kills*, will talk on

"What the Ozone Layer Means to You." There's no admission charge, and everyone is invited.

The Bucks County Environmental Association was formed to study, and try to prepare for, the challenges to our world through such things as global warming, ozone destruction and pollution. As Alice Jones, President of the organization, says, "If we want our grandchildren to have a decent world to live in, we have to start protecting it for them now."

Reporters welcome.

For further information call 555–0000.

That's the basic form. You type it up, take it to your local copying shop, have them make up as many copies as you need (the price will probably be a few cents each). Some copy shops will put your job on recycled paper at no extra charge if you ask for it; if you can find one that does, do it—and add a line at the bottom to say "This release printed on recycled paper" to show you're serious. (And if you can't find a copy shop that can do that, then merely by asking you encourage them to *start*.) Then just put the releases in envelopes and mail them out.

Try to time the mailing so that the releases will arrive in time to be published a few days before the meeting. For a daily newspaper or a broadcaster, you should generally mail no more than a week ahead. The phone number you put on the release, of course, belongs to one of your members who volunteers for the job and is likely to be home most of the time to answer the telephone—or has an answering machine, at least.

Will the papers publish your release? That is not guaranteed. Sometimes they will, sometimes they won't; in either
case, keep sending them out anyway. The smaller the medium,
the more likely they are to print your release, call for additional information or even send a reporter to cover your
meeting; but even the large city papers will sooner or later
pay some attention to you anyhow. They just won't do it as
often or as reliably. If you keep on sending your releases out,
at some point in time there will come a slow news day when
they'll have nothing more urgent than your story to fill the
paper, and then you're in.

The radio and TV stations are chancier. Apart from the
community cable stations (which often simply run lists of
community activities), broadcasters don't usually have any
standard format that would include space for such stories . . .
but they may well send someone to cover the meeting (possibly with a microphone, or even a whole camera crew), or
even invite one of your members to appear in a sixty-second
spot on the news broadcast. Thus you become a television
star overnight.

Don't limit yourself to notices of meetings. Send out a press
release any time the club does any sort of public activity—
whatever that activity is: sending a delegation to visit your
Congressperson, appearing at a zoning board meeting, etc.

Newsletter

And, between times, you should consider publishing a newsletter.

The newsletter is the way you keep all your contacts aware
that you're still alive. You should send it, at least, to all of your

members, including the inactive members, and to all the people you ever considered prospects for membership. You might well want to send it to others as well—perhaps to legislators, or to someone in a legislator's office who has seemed at all interested; to news media people who have shown an interest, if there are any; to community leaders and anyone else you can think of who might care.

In this era of desktop publishing through computers, your newsletter can be as fancy and professional-looking as you like. It doesn't have to be fancy, though. A couple of letter-sized pages, mimeographed or duplicated at your copy shop, will do the job.

What do you put in the newsletter?

First, you say what happened at the last meeting, and what's going to happen at the next. (Thus making it do double duty by serving as a meeting reminder for your members.) You mention what the activities of the club have been. If there have been interesting local developments about landfills or pollution or whatever, you recap them for your members. You suggest letters they can write—if a vote is coming up in Congress on an environmental bill, to their Congressperson and Senators, etc. If any of your members has done anything interesting, tell about it.

You may also want to personalize the newsletter by listing members' birthdays or wedding anniversaries, children's birthdays and graduations, etc. You may want to recommend recent books or articles. You may want to quote some interesting authority with an interesting line or two on some environmental question. All this sort of thing is called "filler." It makes the newsletter more readable, and you can run as much of it as you have room for.

These activities require spending money, of course. So far we haven't been talking about a whole *lot* of money, but maybe even the amounts required for these activities will come to more than you can easily get through collecting dues or passing the hat. Which brings us to your next club activity:

Fund-raising

This is not a fun part of running an organization. Still it is a sad fact of life that unless you do some of it you won't have much of an organization to run.

How much fund-raising you have to do depends on what you need to spend. To find that out, it helps to start by preparing a kind of loose budget.

You will probably find that among the expense items you will need money to rent a room, to buy coffee and doughnuts for after the meeting, to pay the copy shop for your publicity releases and newsletter and to buy stamps.

You may be able to escape some of those costs. Maybe you can get the room for nothing, and maybe somebody will donate the coffee and doughnuts; but some money you will definitely have to have. Make the budget as complete as you can to reflect your actual needs.

If your projected actual cash expenses come to fifty dollars a month or less—enough, say, for the news releases and the newsletter—you might be able to finance them out of dues and donations. If you need more, you're probably going to have to scrounge for it.

Among the most popular of the usual fund-raisers (the legal ones, anyway) are things like dinner-dances. These can be of any magnitude you think you can afford. The local

Holiday Inn or wedding-reception palace will be glad to quote you prices on a room and a meal for whatever number of people you think you can attract. Then you take their cost, plus whatever other expenses are involved (music? programs? cost of printing tickets and advertising flyers?). You add a few dollars a head for your club's profit and start selling tickets.

When this works, it's fairly painless and can even be fun. What's wrong with that is that it's risky. You can actually lose serious amounts of money on such an affair if you don't get the turnout you need. You will have to guarantee a certain number of diners to the hotel or caterers, and they will seldom be sympathetic if your ticket sales don't make the guarantee. That's bad news, for when the group works hard and still winds up with an out-of-pocket loss the event is bound to have a very depressing effect on both the club's finances and its morale.

So exercise caution. If you are lucky enough to have a few members with experience in running this sort of affair, and if they are convinced that an enterprise on this scale is feasible, why, then go for it. If not, play it safe and start small.

"Small" can be quite small. If you have a few members who fancy themselves as cooks, and are willing to go to the trouble, each of them can make some money for the club by giving a cash-on-the-line dinner party. Many organizations have done this very successfully for at least small-scale fund-raising. All it takes is for each of the volunteers (or more likely volunteer couples) to invite three or four other couples for an evening, cook up a storm, charge the guests a "donation" (calling it that in the hope of avoiding tax troubles) of some reasonable amount, say about what they would pay for a decent restaurant meal. The surplus goes into your treasury.

Larger, but still manageable, would be a covered-dish dinner dance, *not* catered professionally. Each couple attending brings a casserole, a dessert, some appetizers—whatever. You need a place to have it, of course. In summer, perhaps someone's backyard; otherwise search around until you can find one (preferably with some sort of kitchen facilities) that's cheap enough. Or your members (and their guests) might like to have an ecologically sound dinner—range-fed beef hamburgers on an outdoor grill; or even a vegetarian meal, to relieve the pressure on agriculture by short-circuiting the grain-cattle-steak food chain. However you do it, you'll probably want to have music. Records will do.

Dancing does not have to be to amplified rock or golden oldies from the big-band era. If you can find a square-dance caller, that's often fun, and good exercise, too. You may have a folk-dance fan in the group—they seem to go with ecology types—and with him to teach a few of the steps and lead the dances, that could be even better.

Another revenue source, ecologically very sound (but not likely to produce big bucks), would be to collect deposit bottles or aluminum cans and turn them in for cash. If your community is already recycling that might not be a good idea, because of interference with the municipal program. On the other hand, if it hasn't begun any recycling program at all, you might be able to collect and sell wastepaper and scrap of various kinds. The money won't be likely to be worth all the physical effort required . . . but look what you'll be doing for the environment at the same time.

You can sell things. You might want to run a bookstore

for the club, taking orders for ecological (or other) books, getting them from the local book wholesaler at a discount and selling them at list price. The downside of this one is that it probably risks getting in trouble with the authorities unless you're willing to go to the trouble of getting a permit, collecting sales tax, etc. (But non-profit organizations can often be exempted from some of the red tape; here too, as so often elsewhere, a friendly lawyer can give you good advice.) Or you can have a cake sale; or a garage sale where members donate things to be sold. Or you can sell advertising space in the newsletter to local merchants. That has the drawback that someone has to go around and solicit the advertising, which is not usually a lot of fun. Merchants generally hate it when their customers hit them up for twenty dollars or so for such things. Still, most of them are philosophical about it as a normal expense of doing business, and whatever you take in in that way is pretty nearly pure profit.

Less lawful fund-raisers (but commonly done) include things like raffling off small "door prizes" at meetings, etc. If you want to go big scale, you can raffle off something really substantial—a trip for two to Hawaii, a new car, a weekend in a nearby hotel—but for that you do need a license; you can't afford to take too many chances with the law on a conspicuous lottery. There's a still worse risk. Although you can often get quite decent prizes for relatively low prices from friendly merchants, there too you run the risk of losing significant money if you don't sell enough tickets.

If you can involve some willing teenagers to enslave themselves for the cause, you can sell off their time for car-washing, baby-sitting, lawn-cutting or whatever else they're willing to volunteer for.

Or you can simply ask for donations. How well this works depends largely on how affluent your group is; a few people who will throw in a hundred dollars or so apiece can save a lot of boring effort.

In any case, do your best to spend as little time as possible on fund-raising—because it's boring, and because you don't want to burden your members any more than you absolutely must.

Action at Last

So far we've been talking mostly about survival matters, the things you have to do just to keep the club going, and relatively simple forms of reaching out to influence the outside world.

If things go at all well, though, you'll want to do more than that; you'll want to *accomplish* something that will make the world better. That's what we'll discuss now, and we can call this section:

Practical Action and Consciousness Raising

What you can do in this area is limited only by your ingenuity and your willingness to put in the necessary effort. *Everything* you do will have a double effect—not only to accomplish its ostensible purpose, whatever that may be, but also to make the neighborhood know you're around and to remind them that serious actions must be taken if the world is to be kept livable.

What specific challenges should you undertake? An easy beginning would be to tackle at least the fringes of the problem of plastic waste.

Your supermarket probably likes to bag your groceries in plastic, because it's convenient for them. Technically this stuff is recyclable, but that's an illusion. In practice hardly anybody recycles it, so it continues to clog up the landfills. But it needn't. Shoppers can reduce that extra load of polluting plastic by saving the plastic bags when they get home, carrying them back to the supermarket on the next trip (they weigh almost nothing) and using the same bags over and over again.

A good consciousness-raising effort in this area is to find a volunteer, set up a card table in front of a supermarket and hand out leaflets asking customers to reuse (rather than try to recycle) their plastic bags.

You may feel a little diffident about this, but remember that this sort of public expression of your views is among your freedoms of speech guaranteed under the Constitution. The store manager probably knows this and won't complain. (After the first shock he probably won't complain much in any case, since in the long run this will save him the cost of some number of extra plastic bags.)

Will this accomplish anything? Yes. Some of the customers will do as you suggest, and some pollution will be avoided. Most of them probably won't; but among those who do a few will love the idea, and, since you have put the name and address of your organization at the bottom of the leaflet, they may well be your next generation of recruits.

Under those same Constitutional provisions, you have the right to put up a card table in just about any public place you like for the purpose of getting signatures to petitions (to legislators, to park commissions, to just about anybody whose actions you would like to affect).

The kind of place you want to do this sort of thing is where there is a fair amount of traffic, and there may be some question about which places of this sort are legally "public." Airports have been generally declared public, due largely to the efforts of groups like the Moonies; shopping malls pretty generally demand to be considered private. But if the proprietors of the place you have chosen won't let you set up indoors, they may not object to the use of a space in their parking lot. (It's worth discussing the matter with the management before doing anything. They may give you permission to come inside, or at least suggest a satisfactory outside place.)

In fact, it's worth sending a delegation to the management of any enterprise that figures in any way in your plans.

If a local factory is still discharging CFCs into the air, or allowing toxic chemicals in its wastewater runoff, a couple of your members can phone to make an appointment, drop by and talk to them. Don't tell them they're villains—at least not right away. Simply ask them what plans they have to change, and when they plan to put them into effect.

For more conspicuous public events, you might want to stage a few demonstrations. On trash: get some household to save up a week's trash and mount it in a parking lot somewhere to show just how many pounds of paper, glass, plastic, metals, yard waste and simple garbage each of us produces each week, and how the quantity can be reduced through curtailing excess packaging, recycling and mulching, etc. (Be sure to let the newspapers know about that one.) Or you may be able to produce some ugly-looking samples of pollution from a local factory or utility, and display a few jars of them in some public place. You might even want to picket some serious and unregenerate offender, at least long enough to let the local paper take some pictures of you doing it. Picketing is another of those Constitutional rights for which you don't need anyone's permission, but it wouldn't hurt to have a friendly talk with your local police desk sergeant ahead of time to set up ground rules about when and where you do it.

It's worthwhile to set up an environmental monitoring service, so that one or more members of your group sit in on all public meetings of local zoning commissions, health boards or whatever other body has a bearing on the ecology. Your watchbirds can do two things while they're there: They can listen to what goes on, and report to the members about it at the next meeting; and they can stand up and speak their piece or even testify, when the occasion is right.

Finally, you can send a delegation to visit your Congressperson (and anyone in a similar position) to ask them where they stand on all the ecological issues, and to try to urge them to stand right.

Remember, you don't have to do any of this *alone*. The national and worldwide organizations do this sort of thing all the time, and would be delighted to have help—or even to provide it.

If your club doesn't wish to affiliate with larger organizations, at least some individual members may want to join them. Many of the national environmental groups will be glad to provide advice, and possibly will even send speakers to address your meetings as well.

How are you bearing up under all this so far?

Some of the things we've proposed may well be causing you to have second thoughts. As a decently reserved person, throwing your weight around in these public ways may not be your very favorite thing to do.

You may not look forward to confrontations with factory managers or shopping-mall owners. You may not want to upset your local supermarket manager, either. You know that he has worries of his own (keeping the volume up, reducing spoilage in the fresh vegetables, watching out for shoplifters, making sure he has enough minimum-wage workers showing up for work to bag your groceries). Is it fair to add to them?

But the answer to that is clearly "yes." His business practices are adding to the problems, aren't they? And once he changes you'll leave him alone.

You may even be a little apprehensive about what effects doing this sort of thing may have on the quiet pursuit of your normal, inconspicuous life.

Most Americans have apprehensions of that sort lurking somewhere in their minds. They seem to turtle up in regard to their government, pulling in their heads and limbs out of some vague, nameless fear that if they call the government's attention to them the government will make them wish they hadn't.

That is really not at all likely to happen. Although there are all too many officials who wish you hadn't, you still do have all your civil rights. The law protects them, all the way up to even the present Supreme Court. The local officials aren't legally permitted to punish you by putting you on jury duty, nor will the Feds be any more likely to call up the Internal Revenue Service and have them bring you in for a tax audit because they're mad at you—after all, Richard Nixon isn't the President of this country any more.

So don't neglect the task of keeping your Congressperson (and others) aware of your existence, and letting them know exactly what it is that you want them to do.

You may have been doing that already by mail. Even more effectively, you can do it in person as well.

When you have a decent number of people showing up at meetings (how many is "decent"? The more the better, but as few as a dozen warm bodies might be enough), you might even invite your local Congressperson to come and talk to you about some specific problem.

He may not come himself, especially if he's pretty sure that you haven't been supporting him or his record. Still, sooner or later, if you have convinced him that you represent some significant number of voters, he is likely at least to send some staffer from his office to represent him, and that would be almost as good.

See, the reason Congresspeople, like all public officials, are where they are is clearly spelled out. It is for the purpose of doing what is best for the public. You're members of the public. You have every right to let them know what you think is best.

Whether they will listen to you, on the other hand, depends a great deal on who you are, and on what kind of politicians they are.

So let's take a break from the how-to-do-it plans for a moment, and look at just what politicians are really like, and what it takes to get them to do the right thing.

23

Politicians

If you ever want to get anything done on the government level, the people you will generally have to get to do it for you are either bureaucrats or elected officials. Whichever they are there is a name for their sort of person. They are called "politicians."

We all know what politicians are. They are the silly goblins in the newspaper's editorial cartoons, and the subjects of an endless barrage of jokes from every stand-up comedian in America. Everybody else jokes about them, too, for their venality and stupidity and their very well publicized nasty personal habits.

But who is the joke on?

Answer: Well, who put them in office?

It is sadly true that it is very difficult for any politician in America to be squeaky clean. We make it hard for them. We give them contradictory commands.

First, we expect them to obey the charges laid on them by the Constitution. Under that document, their duties are clear. Collectively the 435 members of the House of Representatives and the 100 senators are given the task of passing all the laws which govern our country as a whole. That means that they are the ones in all the country who are entitled to write the legislation which will "provide for the common defense, promote the general welfare and secure the blessings of liberty to ourselves and our posterity"—as the rules are laid out in the Preamble to our Constitution.

However, we have laid another obligation on them as well, and it goes in the other direction.

Each individual congressperson is also required to "represent" those particular voters who elected him.

That means they are not only supposed to promote the *general* welfare—that is, the welfare of the nation as a whole—but at the same time to take care of the *particular* welfare of their own states or districts. This they generally do by passing out money. Tax money. *Our* money. They do it by making sure that that little slice of the nation which is their own turf gets as good a deal as possible when tax money is being spent or tax breaks are available, even when that has to be at the expense of other areas.

Those two objectives are not only in conflict, they are almost always completely incompatible.

For instance, when it comes to deciding which particular community is going to get the benefit of a new government agency, airfield or naval homeport, with all the jobs and wealth that flow from such enterprises, every last congressperson in Washington covets it for his own. If he is successful in getting it located there, his district profits. Other districts lose out.

That causes problems for the representatives of the other districts the next time an election comes along, since their "constituents" (which should mean their voters, but as we will see all too often means something quite different) also want a piece of the pie.

Congresspeople as a class are pretty tender toward their colleagues' problem, even when the colleagues are also their political opponents. To keep any of the legislators from being too badly hurt, they will generally create more such enterprises than the general welfare really needs—however useless the enterprises are—so that everybody can get a taste of the loot.

This is colloquially called the "pork barrel."

Pork-barrel politics is the sole reason why the United States is afflicted with so many outdated, useless and expensive military installations—not to mention subsidies, federal grants and other money sinks—which go on draining countless billions of dollars out of the national budget (that is, out of the money we all pay in taxes) every hour of our lives. It is why important programs never do get funded, because some politician has already siphoned off so much money in pork that there's not enough left for what would indeed promote the general welfare. It is why the federal debt has risen into the hopeless trillions of dollars, and why our grandchildren's grandchildren will still be paying off the money our recent governments have borrowed for this spending spree.

Spending wasted money on boondoggling projects isn't the only way our presidents and congresses harm the "general welfare" they are supposed to protect. Besides the pork barrel, there are many other ways in which a congressperson can enrich his district (or a president his political supporters), at the expense of the nation at large.

One of these ways is bending the laws to favor their own local special interests.

This is a real problem for environmentalists. As we have pointed out, there is no way for us to make any real headway in cleaning up the environment without causing a lot of financial stress. The stresses won't fall equally on all. Some districts are certainly going to suffer more than others. The representatives of Iowa farmers or Texas oil and cattle people may clearly perceive that it is necessary for the country (and the world) to sharply reduce logging in the Northwest—but the representatives from Oregon, Washington and Alaska will suffer great pain from the jobs lost there in the process. And the minute a bill to achieve that sort of result is introduced, every one of the legislators whose constituents are at risk will find his mailbox full, his telephone jammed and his waiting room packed with people who want to tell him why that pernicious and uncalled-for legislation will destroy their livelihoods.

Most of the people who are complaining the loudest will be no strangers to your congressperson.

Some are his own campaign workers. Some are people he grew up with. And a great many of the most insistent, it is sad to say, are known to him in less benign ways: they give him money. They are, in short, his *real* "constituents," and they get in to see him, and what they want is what he tries hardest to give, when the ordinary average-citizen voter in his district never gets past the receptionist at the door.

We're not talking about outright corruption here (although the sorrowful history of the last two hundred years tells us there have been plenty of elected officials who have sold their

votes for cash on the line). Even if your congressperson is in
at least that sense scrupulously honest—honest enough to refuse an envelope stuffed with hundred-dollar bills when it is
slipped into his jacket pocket—he is very unlikely to refuse
everything. He will recognize a lot of the people urging him
to enhance their own prosperity, at whatever expense to the
general good, as the very same people his campaign committee has been tapping for large contributions to his re-election
fund.

He may have other kinds of favors to repay, as well. He
may well have spent weekends at their resort hotels, flown
there in their company planes; he has very possibly given
speeches at their dinners for large honoraria. At the very
least, these wealthy special pleaders are sure to have powerful friends in his own community. The congressperson finds
it very hard not to listen when they come to him with their
arguments. Especially when the arguments are at least super-
ficially plausible. They will be made to sound plausible, too,
because the hired-gun experts the special interests hire will
generally manage to make them sound that way by massag-
ing the facts until they come out right.

The main argument your congressperson will hear against
almost any significant environmental action is always economic: this measure will cost money and *jobs*.

This is not an argument most politicians will ignore. It is
the main reason why George Bush, who campaigned as "the
environmental president," did so little for the environment
once he was elected. He had to choose between protecting
the environment and encouraging economic prosperity. The
environment lost.

Here, too, Bush was responding to pressures that go beyond any particular polluter or despoiler. It is not just the lobbyists who move him here, for he gets that pressure from all of us; presidents do not get reelected if the economy went sour on their watch, whatever the reasons. The American people, and especially American politicians, are addicted to growth. When environmental concerns threaten to slow down that economic growth passions rise.

So the special pleaders will tell your congressperson things like, "My oil refinery will have to be shut down, and there go the jobs of 1800 of your voters." Or, "If my electronics assembly plant can't use CFCs, we'll lose orders to foreign manufacturers who do use them and we'll be out of business." Or, "My trucking fleet can't afford the extra cost, so please exempt diesel fuel from the bill."

They won't stop there. If the pressures of the real world are so strong that the petitioners can't hope to have environmental measures abandoned entirely, they'll do the next best things. They'll ask for specific exemptions for themselves; if that doesn't work, they'll simply play for time. "Maybe some of these environmental horrors won't happen, after all—scientists can be wrong, can't they? So let's not rush into anything; let's spend a few million bucks on some five-year research programs to make *sure* these very radical actions are absolutely necessary before we actually *do* anything."

Unfortunately, these businessmen, paid lobbyists, hired "experts" and Political Action Committees don't stop with presenting arguments. They also carry a club.

If a congressperson is obdurate to their urgings, they have ways of getting his attention. The club they carry is their money, and they will use it on him to do him harm. They may spend it on advertising and TV commercials to whip up popular support for their demands—letters from constituents, picketing, marches.

Of course, you realize these are the very same things you are doing. The big difference is that there will be a lot more of them from the special pleaders, because they are a lot better financed than any environmentalists. If worst comes to worst, and a legislator has the strength to resist their pressure, they'll get rid of that legislator. They will find a more pliable candidate to run against him the next time around, and the campaign contributions that helped put him in office will now go to his opponent.

It takes a pretty strong-minded legislator to resist this kind of pressure. We don't have many legislators like that—or Presidents, either.

That is what you are up against. You have all this tremendous force applied to your congressperson from the other side, all the time.

If you don't want him to give in to it, there's only one thing you can do. You must put some pressure on from *your* side—which is to say, the world's side, and particularly your grandchildren's side.

But maybe your congressperson isn't going to change no matter what you do. In that case you may have to go a bit farther.

There is a way to get environmental laws passed without much help from politicians. It's rather complicated, but it can work even if almost every legislator in your state has been bought and paid for by the polluters. It's called the referendum.

A referendum is an election like any other election, except

that you don't vote for people. You vote for (or against) a principle. In California this sort of referendum measure is called a proposition; in other states it's called a public question; whatever it's called it is an expression of the will of the people. Sometimes the measure is binding, and then it becomes law without any further action. Sometimes it's nonbinding, which means that the legislators can then ignore it if they want to—but will no longer have the excuse that their constituents—their *real* constituents, not the ones who kick in money to their campaign funds—don't really want it, because the evidence is clear.

If this is so easy, then why bother with politicians at all?

In the first place, it isn't easy. To accomplish anything significant a referendum pretty much has to cover an entire state. No little conservation club can manage that. It takes a large and active statewide organization, if not a number of them working together, to have any hope of success; and before you start you need to have established a climate of public opinion that, by every measure you can find, is solidly on your side—which means that all those groups will have to have done a lot of preliminary publicity and general spadework.

Even then it may not work.

California conservationists thought they had the odds on their side when they managed to get Proposition 128—familiarly known as "Big Green"—on the ballot in 1990. The polls showed a majority of voters agreed with its far-reaching provisions, which included measures to save the ozone layer and the redwoods and just about everything else that was threatened in the state. They had even accumulated a campaign fund of $5 million to see it through . . . and they lost.

Partly they lost because the polluters outspent them—*they* threw more than twice as much into the campaign. Partly they lost because a war had come along—President Bush was beating the drums for the attack on Saddam Hussein's Iraq—and so had a recession, and a lot of people who had been concerned about the environment were now suddenly more concerned about jobs. At another time "Big Green" would probably have won, in spite of all the money spent to oppose it . . . but the time went wrong.

So the referendum route is possible; but if you try it you have to recognize that you may lose.

Then you're back to your congressperson.

He certainly won't be any more amenable to your point of view after a referendum defeat, so then you may have to go a little farther: throw him out and elect someone else in his place.

That's not outside the bounds of possibility, either. The next chapter tells you how to go about getting the votes out to accomplish it.

"Getting out the vote" doesn't sound like any fun at all. It's what you expect a Chicago sanitation worker to have to do in order to keep his city job, or a budding lawyer to do for the sake of advancing his career.

It's not all that bad. But even if it were more boring than filling out your tax return and as painful as a root-canal, it is still worth doing.

It *works*.

In particular, you might be reassured to know, the specific how-to-do-it steps that we are about to propose are known to work. We know that, because we've already seen them work. Many of the suggestions on organizing groups and running elections are drawn from a book one of us (Frederik Pohl) published twenty-odd years ago, called *Practical Politics* and based on his own experience as a county committeeman in the 1960s.

The instructions the book contained for running a campaign worked then. They still do work.

The proof is in what some of the book's readers reported. Shortly after *Practical Politics* was published, a citizens' reform group in the city of Cape May, New Jersey, got hold of a copy, read it and did what it said to do in running a campaign. They won their election, electing a reform mayor and a majority of their city council. In Georgia, in the suburbs of Toronto, Canada, and elsewhere in North America other groups had similar success. And, though the book has been out of print for years (because changes in party organization outmoded some parts of it) and was never distributed outside the United States in the first place, there are even recent success stories from a good deal farther away.

In 1989, in Moscow, USSR, a young man named Boris Kagarlitski was about to experience the first free election ever held in his lifetime. He wanted to run for office. He had the motivation to do it (he had been a political prisoner himself, in the pre-perestroika days), and he certainly had the desire. But he didn't know how to go about it. Outside of the movies, he had never seen it done.

Of course, no one else in the Soviet Union had any experience of such things, either. The USSR had not experienced a contested election for more than seventy years. Kagarlitski, however, had an idea of where to go for help. He knew that his father, Professor Yuli Kagarlitski, happened still to have in his library an old copy of *Practical Politics,* given to him as a literary curiosity in the course of a lecture tour when it came out.

So Boris borrowed his father's book. He translated the relevant parts into Russian, handed them out to his political allies, followed the instructions it contained—

And then, when the election was over, he telephoned to say: "It worked! I've just won the election. Now I'm a Deputy in the Moscow City Soviet, and I owe it to you!"

If following these directions can win elections from New Jersey to Moscow, the chances are pretty good that they can work for you, too.

Of course, how-to-do-it instructions are not all you need to win. That's all right, though, because both you and Boris Kagarlitski have something else on your side.

Just going through the motions of the vote-getting process isn't necessarily enough to win an election—though they are what have worked for a thousand political machines in America for generations. You probably can't win without following them, but it's also a great advantage to your hopes of winning if you have a cause that people can see a reason to rally around.

You have that. You have the preservation of our wonderful world for your cause, and who do you know who has a better one?

24

Green Politics

Actually if you've been doing all the kinds of things we've talked about in the previous chapters, you've already made a really good start on winning elections. All you have to do is convert the group you've already organized into a political club.

A *political club*?

Please don't recoil in horror. We know how you feel. We understand that you can ask the average intelligent, well-meaning American to do many things for the sake of saving the environment, but there's a limit. Getting into the dirty business of politics doesn't sound any better to him than suggesting he get down into the sewers and muck them out with his bare hands.

But, you know, politics doesn't have to be that way. Why is there so much corruption in American politics? Because good citizens like us stay out of it. We leave politics in the hands of the bosses and the patronage dispensers. A great many of us may even refuse to vote in a primary election, because we don't want to sully our principles by declaring allegiance to one particular party. So we proclaim our lofty independence.

The political pros of both parties are grateful to us for our principles. They love "independent" voters. Independent voters make it so much easier for the pros to keep control of the party machinery . . . and for the professional politico the party machinery is what politics is all about. That's where the power is. Therefore that's where they can keep firm control of all the jobs, the patronage, the *money*.

Don't forget about the money part. As is true in every line of endeavor, what distinguishes a professional politician from an amateur is that the professional makes his living out of it. That living can be very high.

We don't have to be so obliging to the pros with our vaunted "independence." We can go after a piece of that power ourselves. Once we have it we can have something to say about who the candidates are, and what issues are debated.

Politics isn't a Skull & Bones Club. You don't have to be invited in. Entry into the political system is open to all of us.

As a matter of fact, if you are willing to put the necessary work involved—and not all that much of it, either—you can actually become a political boss yourself. What's more, you can astonish all the other bosses by being idealistic, unbribable and *honest*.

The average well-meaning American voter thinks he has done his part for his country if he stops off at the local polling place the first Tuesday in November and marks up a ballot. In his innocence, it does not occur to him that by that time the action is just about over.

By then all the candidates have been chosen. About the only decision that is left for the voter to make is whether to vote for Tweedledee or Tweedledum. (Actually, there may not be anyone on the ballot he is really *for*, so what he usually decides is which one to vote *against*.)

By Election Day the issues have been decided (or, more often, the various contenders have agreed to ignore the really important ones). The voter may think that all good citizenship demands of him is that he decide which are the good candidates and then cast his vote for them in the general election, but what does he do if by November there aren't any good candidates left?

To have the chance to vote for a good candidate, you have to be around when the candidates are chosen.

What that means under the American system is that if you want to have any say in who will be running in November you therefore have to get involved in the affairs of a political party. As a minimum, you have to vote in that party's primary election.

Some unsophisticated voters are reluctant to declare for any party at the time of a primary because they think it takes away their freedom of choice. They don't want to be compelled to vote for that party's candidates when the general election comes around.

Well, of course that's silly. They're not compelled. Figure it out for yourself: they *can't* be compelled, because that's impossible. After all, the ballot is *secret*, so no one could check up on you to see that you voted "right." And there's nothing in the laws that requires you to do that, anyway.

To be really effective, you have to go farther than just casting a primary vote. For that you need to become *active* in the party—even to the point of holding office in it.

Assuming you are interested in going that far, there is a

decision that you have to make right up front: Which political party do you choose?

Use your own discretion about that. If you don't have any better way of deciding, flip a coin. The truth is that there isn't much reason to worry over which way you choose, because when you come right down to it there's not really a great deal to choose between them. For our purposes, there are liberals and conservatives in both parties who are simply not interested in most environmental matters—and indeed may even be hostile to them, because they threaten other interests—and there are a few of both stripes who are quite sensitive to them.

Here, too, the fact that the decision is hard doesn't excuse you from making one. You can't cast a "nonpartisan" ballot in a primary election, nor can you have much effect on the inner workings of any political party without taking part in them.

As a practical matter, your choice at this time in American history is between either the Republicans or the Democrats, because they are the parties which are likely to succeed in electing candidates. Consult your own emotional preferences, however slight. The important thing is to remember that, if you want to accomplish anything, what you *can't* choose is to be neither.

Having picked your party, how do you bend it to your will, at least to the point of choosing good candidates and dealing meaningfully with the issues of the environment?

To answer that, we need a short civics lesson on how party organizations are constructed.

Between elections, the affairs of both political parties are run by standing committees.

There is a national committee, made up of delegates from each of the fifty states and a few districts and possessions. These people are mostly dedicated party workers who have been around for a long time; they are elected by professional politicians from their own states and it will be a while before many serious environmentalists make it to that level.

Below them are the state committees of each party, which are also mostly professionals, and below them still are the county (or parish) and municipal committees for each party. At that low level—in fact, the absolute lowest elected office anywhere in America—a few of the committeepersons will be professional politicians. Most of them will be part-timers who collect a political favor now and then. And there will even be a few who do it for fun—or for principle.

Think carefully about those municipal committees.

There, at that lowest of all levels in a party's administration, is where you—or any other member of your club, possibly several of them—can most easily find your entry point into the workings of the machine.

The exact party structure will vary somewhat from state to state. (To find out what yours is like, get on the phone again; call your county clerk's office and ask them to spell it out for you.) Typically, the county committee is made up of one or two persons elected in the party primary each year from each election district within the county; and the municipal committees are local factions of the county organization.

We said that was your easiest entry point. We should warn that that may not be true in your part of the country.

In places where a party political machine is strong, rich with patronage and well staffed with "volunteers," it is not at all easy for an outsider to get elected to the party's county committee. There may well be contests in the primaries. If there is ever a *serious* struggle for the post, though, it will almost always be a struggle for power between a couple of strong factions within the party. No radical environmental upstarts will be involved. Both sides of the contest will be made up of party stalwarts, skilled and well organized, and they will put a lot of effort into getting the vote out for themselves on primary day.

But that's true only where the party machines are strong and well entrenched. In most of the country, especially outside the major cities, party organization is likely to be a lot flimsier, and there the committee posts are seldom hotly contested. The turnout on primary day is usually small. A handful of votes may be enough to elect.

Surprisingly often, one party or the other may not even be able to find anyone to run for those offices at all, in which case a single write-in vote may be enough to put some of you in office.

For party officials, the primary election is the bottom line; there are no party offices at stake in general elections. And if you are fortunate enough to live in one of those weak districts, this is where your opportunity lies.

You may wonder what value there is in being elected to a weak political machine. The value is that, whether the party is weak or strong, it's still true that the county committees are the people who have the power to choose the state committees. They are also the ones who have most to say about who their party's candidates will be.

True, now in America most candidates are chosen by votes of anyone who cares to show up and declare himself in the primaries. The days when the candidates were hand-picked by the famous pros in the smoke-filled rooms are over— sometimes. But those pros still have enough power to choose a "regular" slate—and then to muster the party workers to get whatever signatures may be necessary to put that slate, or anything else they like, on the ballot, and finally to get the vote out on primary day to bull their choices through.

These county leaders are not put in office by God. They get *elected* to their leadership posts, and the votes that elect them are those of the whole county committee (or sometimes of the municipal leaders, who themselves are elected by the county committees in their own municipalities), and those votes belong to the elected county committeepersons. If you go that road, and become a member of your party's county committee, the only significant thing you may ever do in your formal capacity is vote in the annual organization meeting that follows the primary. But that vote can count for much.

What also counts is that, as a duly elected county committeeperson, you have the right and opportunity to raise your voice in meetings.

This is a precious privilege, because, although you may not be able to dictate a choice of candidates or policies, you have a pretty good chance of *affecting* them.

The reason for that is that most professional politicians have no interest at all in issues.

For most politicians, "issues" are the name you give to the public-relations strategies you use to get voters to like you. The name of the game, for them, is winning elections.

Political machines are not involved in policies, only in power. If you have some of the necessary power, even only as little as your annual vote on the leadership of your county committee, you begin to matter to them. Then they would rather please you than frustrate you, everything else being equal. And the more voters they think you represent, the more seriously they will take what you say.

That's one way to make your little environmental club become a factor in the constantly changing equation of how the politicians run their affairs.

But if you're going to go that far, you might as well go a little farther. You might as well *be* a political club, doing the things that are expected of it.

None of those things is beyond your power. One thing that political clubs do, for instance, is to supply the party organization with some election-day manpower to get the vote out. Election workers include both the people you see hanging around the polls when you vote and the others you don't see because they are at home working the telephones, or they are in their cars, driving the handicapped or the merely lazy to the polls, and thus getting the vote out.

The people who actually work in the polling place itself come in two varieties. One group are the statutorily required election judges and clerks; these have to be there, by law, and they are paid (though not very well) by the state elections board.

The other people who you will see standing around the election place with no evident official work to do are the poll watchers.

They represent the party organization, and they are there because each political party is allowed to name one or more watchers to hang around and make sure no one fools with the voters, the ballots themselves or the count at the end of the day.

They do more than that, of course. They keep an eye on who has voted, to make sure that all the ones they know are on their side have come out. They even have an official duty which is that they have the right to step in and challenge someone's right to vote, if they think there's something wrong with his registration (and especially if they think he'd be voting the wrong way).

These watchers or "challengers," as well as the outside party workers who are doing other tasks, get paid, too, but not by the government. Where a party machine is strong they probably get paid quite handsomely and in advance— by having been given light-labor patronage jobs, knowing when they accept the jobs that part of their work will be extra-curricular on behalf of the party machine. Most of the others have their payday on Election Day itself when, at some time not long before the polls close, the local municipal boss or his representative will come around with an envelope full of money to hand out to the "volunteers" for their pains.

The work these people are doing is called "getting out the vote," and it is how elections are won.

It is important to remember that there are two separate battles to be fought in any election. If you want your side to be declared the winner at the end of the day you have to win both of them.

One you might call "the battle for voters' minds." The other is the battle for their bodies, or at least for the parts of their bodies they vote with.

Which of those battles is the more important? That's like asking which you need to live, your heart or your lungs.

In a political campaign, we environmentalists have to have them both. The political machines don't. As long as they get the right number of votes cast for their slate, they couldn't care less what the voters are thinking about. So we must both persuade as many people as possible that our cause is right, and *also* get as many of them as possible to do the actual voting on election day.

It is pleasant and inspiring to work on men's minds, and pretty distasteful to work on the actual task of moving them from their comfortable living rooms to a bleak polling place, but that's the way it is. You have a right to regret that this drudgery is necessary. You can't leave it out, though. It would certainly be nice if citizens really spent a lot of time thinking about issues, and came out on the right side of them (namely our side, of course), and then got themselves out to cast their ballots.

But some of them won't, and the rules of the game are that in that case whatever they think doesn't matter. If you stay home you don't get a vote.

It is also true that the vote of the wise, motivated, right-thinking voter counts for exactly the same as the vote of the dummy who votes the way the nice fellow on the street who gave him a cigar asked him to. If there are more of the latter than of the former, then the latter's candidate gets elected.

So, however right your cause, you have to get people to step into the booth and pull the little lever.

There are many stimuli that can produce this response on the part of the voter. Some voters will vote your way because they recognize that the environmental policies your candidate stands for will help the world. Some will pull the lever because you slipped them a five-dollar bill. Some will do it because the name of the candidate has so saturated their senses through TV blitzes that he seems like an old friend. Some will do it because some person they like—maybe you—has asked them to do it as a favor.

The thing to remember is that the voting machine doesn't ask the voter why he pulls the lever. If the lever is pulled, the counter clicks around one notch. The person who masses the largest number of clicks is the person who gets in. And that is the Whole of the Law.

Getting the Vote Out: The Campaign

If you want to have the numbers come out right on Election Day, you will want to get started early. The general name for what goes on in this process is "the election campaign."

That's the same word that's used in warfare, and that's appropriate. An election campaign has a close resemblance to a war. Like a war, an election campaign may be won or lost without much regard to its righteousness. The outcome will depend more on questions of logistics and strategy, organization and training—and luck. You can't control the luck, but you can *prepare* for the election, and the sooner you start the better.

The candidates. People are more likely to vote for someone they've actually met, even if the meeting is no more than a quick handshake at a train station. So you want to help your candidates meet as many voters as possible.

Which candidates do you want to help? (You're not obliged

to give equal support to your party's whole ticket, you know. In the real world of politics, hardly anyone does.) First, of course, you want to help the ones who are heart and soul on your side. Since environmentally aware candidates may be scarce on the ground in your area, your next choice is the candidates who are at least better than their opponents in some significant respect—not only because electing them will therefore upgrade the entire political picture, but because if you work for them, and are known to work for them, then they will be a lot more likely to pay attention to what you want them to do when they're in office.

Your first step is to invite the candidates (or some of them) to come and speak to your club. They almost always will, and that's good for both of you—you get the benefit of discussing with them the actions you would like them to take if elected, and they get to expose themselves to voters.

Then you can "walk the candidate around." On some warm Saturday afternoon you can walk them through a neighborhood or a shopping center, introducing them to people you know—or even people you don't know, because you can say nicer things about the candidate than he can about himself. You can set up "coffees" at the homes of your members, where the member's neighbors (to the extent of anywhere from three or four to as many as the house will hold) are invited to meet the candidate and press his flesh. (And, of course, at these coffee gatherings you have the chance to direct the discussions toward the environmental issues which concern you.)

All of this puts the candidate to some extent in your debt. That's a good place to have him. If he is a successful politician, he will probably try to repay you in some way if he can, perhaps even by voting your way on environmental issues now and then.

That may not be easy for him, for when he wins his election he will owe a *lot* of debts to a *lot* of people. It may well be that on some issues he will feel that he owes his big-money campaign contributors or the local large employers more than he owes you, and then you are out of luck. But at worst you will have more access to him than you did before, and if he can't do exactly as you wish at least he will generally try to do some *part* of it.

At the same time, you're getting ready for the nitty-gritty on Election Day. You procure lists of the registered voters in your district (from your party, or from the local town or county clerk). You mark the ones you expect to be on your side, as well as the ones you don't, and you keep those lists for use when Election Day arrives.

Then, a week or two before the actual day, you get on the telephones.

Your main purpose is to say something like, "Hello, I'm your neighbor, John Smith, and I'm working to help get Mary Jones elected to the State Senate because she understands what we have to do about the environment." Some people will want to talk about it, and then you can discuss issues as long as they will listen. (Not a second longer, though. Your telephone squads should be chosen for their sensitivity to the people they talk to, including an awareness of when it's time to say good-bye.) Some will hang up on you—and can hardly be blamed, with the number of sales pitches we all get on the phone every day. But it needs to be done. Not only to carry the message to the voters themselves, but to help you mark your lists to see which voters appear to be on your side.

At the same time, you're continuing with all the things your club has been doing all along—holding meetings, organizing environmental activities, getting out your newsletter and your press releases. Some will naturally tie in to the actual election campaign. Others may not, but you don't stop doing them anyway.

Finally, you start planning how to use the forces at your command on Election Day itself. If you have the chance to appoint poll watchers, you pick them out, and decide which election districts they will cover.

This is a strategic question. If you have a club of 25 active members, and you want to elect a congressperson who may have 200 separate polling places in his district, you obviously can't cover them all. Which ones do you try to cover?

One factor in your decision is to work in the places where you have friends; you'll be more effective where people know you. The other important factor is how you expect the election district to vote. Given a choice of ten possible districts, half of them favorable to your cause and the other half likely to be opposed, the ones you should choose to work on are the favorable ones.

That may sound strange to you—why preach to the already converted, when you could just as well be trying to convert the heathens? But, as we said above, it doesn't matter how many people you get *on your side*, it only matters how many of them *actually vote*. "Working a district" brings out additional votes—but in a hostile district you'll very likely be bringing out people who will vote the wrong way. As long as your manpower is limited, concentrate on working the richest lodes. Leave the hostile ones alone . . . and hope that your opposition will do the same.

Getting the Vote Out: Election Day

When the great day dawns, the party watchers you have assigned to the polls get there early with their prepared lists, and check off each voter as he or she comes in.

The polls will likely open too early for your other workers to do anything yet, so they can sleep a little later. Starting at a reasonable hour, say 9 AM, your telephone squad can begin to call a few of your most favorable voters just to remind them to come out, and to suggest that they bring a neighbor with them when they vote. Later on, perhaps in early afternoon, your poll watchers will start passing along to your command post the names of the favorable voters who haven't shown up yet. Then the phone squads will call to remind them—maybe even to offer transportation, or baby-sitting, or whatever is necessary to get them out. Many people won't vote until they come home from work, so it isn't until 6 or 7 PM that things begin to get critical. Then you double-check the lists, and get after the laggards . . . always, of course, with all the courtesy and consideration possible.

None of this work is beyond the skills of any ordinary human being who is capable of tying his own shoes, yourself and your clubmates certainly included. It is, however, sometimes pretty boring.

But it is essential. It is how elections are won, and that is what you are in politics for.

By the time you've got your feet wet in one or two actual elections you will have learned a lot more than any book can teach you—whether you've won or lost.

In either case, don't stop. That is *important*. It is also a mistake most political reformers are almost certain to make.

When reformers lose an election, they are likely to get discouraged and turn to something more rewarding to do with their time. On the other hand, if they win the election they are likely to think the question is settled once and for all and they can relax.

Not the professionals. Remember that professional politicians are *professionals*. It is how they make their living. A salesman doesn't quit his job after he makes one sale—or, in disgust, after he fails to make one.

It is the same with the pols. Whether a professional politician wins or loses a single election, he will be back there for the next one. It is what he *does*. The professional politician can afford to be quite tolerant of volunteer reformers like ourselves, because he knows that sooner or later they'll give up politics and go back to whatever thing is the primary focus of their lives. He won't, because politics *is* the primary focus of his life.

So there are no three-year enlistments in the war to save the environment. Once you're in it, you might as well figure that you're in it for life.

You might as well set your sights a little higher, too. Once you have enough experience to see how such things are done, you might start thinking about generating your own candidates—members of your own club, for instance.

There are two ways of getting a candidate on a November ballot: as a non-party independent, or through the party machinery.

For an independent candidate, getting *elected* may be hard, but to get on the ballot all you need is to circulate a petition, get enough signatures, file it with the appropriate authority in time to comply with the law—and there you are. To find out what form the petition should take, how many signatures you need and when and where to file, call your county clerk's office.

Nomination by petition is legal all over the United States, but there are traps to watch out for. Whatever the number of signatures required, try to get at least twice that. The signers generally must be registered voters; they must live and vote in the right districts; and their signatures must match their names on the registration lists. They won't all be all of those things. Some people will sign almost anything, and if you're viewed as a serious threat to anyone, that someone will make it his business to examine every signature on your petition to find the ones which are invalid. Inevitably some will be thrown out, and only the ones which survive that sort of scrutiny will count.

That's if you want to put an independent on the ballot. That probably is not your best strategy. The trouble with running as an independent is that you then forfeit all the knee-jerk votes from those voters who have voted all their lives the straight Republican (or Democratic) ticket, and don't really care who the candidates are. Your chances of election are generally better if you are the official candidate of a major party.

That's not impossible to get, either. If your club has been a force in one of the parties, you're entitled to claim a spot or two on the ballot for the people you want to elect. You may not get it, but then you just go in for a primary fight—which involves the same routine of petition-signing and filing as running as an independent.

So go for it. Even if you lose, you've gained some of those valuable commodities for your cause, publicity and experience.

And if you win—why then, some member of your group (why not you yourself?) can actually *make* some of those important legislative decisions.

25

Is It All Worthwhile?

Let's—reluctantly—come back to present time.

All those good things haven't happened yet. We don't have a majority in Congress, we don't have a president who puts the environment ahead of business as usual, we haven't even yet succeeded in getting all the CFCs and their slightly less destructive cousins banned around the world. Our planet is still in bad trouble, and if we want it saved we've still got the whole job all to do.

Is it really worth the trouble? To put it differently, what do we get out of it all?

That's the best news of all—the first piece of unalloyed good news we've been able to offer you. We, and our grandchildren, will get a *lot*. If we can stop looting our environment and change our way of life to a sustainable, steady-state existence the future is *golden*.

That doesn't mean "perfect." (Maybe perfection is too much to ask this side of heaven.) We can't even promise that the environment itself will be perfect. There will still be violent storms (though, with any luck, not as many, and not as

lethal). There will still be cases of severe sunburn for beach-
goers (though fewer blindings from cataracts, and fewer
deaths from skin cancer). There will still be earthquakes and
other natural disasters . . . and, of course, saving the environment isn't enough to bring Utopia anyway. No matter what
we do for our natural resources, environmental measures alone
can't solve such numbing social problems as drugs, crime,
warfare, terrorism and poverty.

But there is so much that we will have!

If misused technologies have brought us to the brink of
disaster, there are other technologies which can make human
life richer and better than any generation of human history has had since Homo sapiens sapiens first appeared on
Earth.

Medicine will ease nearly all the debilitating and deadly
diseases that blight so many lives and cut off so many of us
in our prime. The computer has just begun to make its effects
felt. Before long few human beings will ever again have to
work at dangerous or stultifying jobs, because smart machines will do the jobs for them. New technologies in communications, transportation and a hundred other areas are
on the horizon, and collectively they will lead us to a miraculously more wonderful way of life . . . if we can only keep
our world intact long enough to reach them.

We can do it, you know.

We can get there. We can have it all. The Third Millennium A.D. can be the green millennium, the time in which
we learn to live as responsible human beings at last.

There is no law, natural or divine, which demands that the
world we live in become poorer, harsher and more danger-

ous. If it continues to become that way, it is only because we do it ourselves. . . .

And to put an end to the practices which are bringing about this decline, the only thing we have to do is to stop doing them.

APPENDIX 1
Sources and Resources

Since this book is written for a lay audience we have not followed the academic practice of citation of specific sources in every case. The main sources were currently published periodicals, primarily scientific and environmental magazines (the most important of which, in the area of science, were *Nature, New Scientist, Science* and *Scientific American;* in the area of conservation, *The Amicus Journal, Audubon, Greenpeace* and *Sierra* as well as *State of the World: 1991*—all of which are described below) and, in particular for details of local or regional events, a large number of newspaper accounts from both the United States and the United Kingdom. Some individuals must be thanked as well. These include Grahame Leman, who was assiduous in combing British publications for relevant material, and Ian Ballantine, who performed the same service for otherwise overlooked American ones. In addition, Carl Sagan, Hans Moravec, Freeman J. Dyson, and John Gribbin provided special insights in personal communications.

The four indispensable periodicals of general science included two English and two American. These magazines continue to provide excellent coverage of the scientific aspects of environmental concerns (as well as scientific information in all other fields) and are recommended to persons who wish to keep their knowledge up to date. They are:

Nature (4 Crinan Street, London N1 9XW, United Kingdom) is one of the world's two leading weeklies of general scientific information. Much of its content consists of actual research reports in all scientific disciplines and may often be too technical for a general audience; it is read primarily by working scientists interested in knowing what is going on in scientific areas other than their own specialty.

New Scientist (110 High Holborn, London WC1V 6EU, United Kingdom) is far more popularly written than *Nature* and is designed for an intelligent lay audience. Each weekly issue generally contains a number of review pieces which explain some technical matters—sometimes in cartoon form. Unlike most scientific periodicals it contains a good deal of humor—unfortunately not always entirely intelligible to an American audience.

Science (published by the American Association for the Advancement of Science, 1200 New York Avenue NW, Washington, DC 20005) is the American weekly equivalent of *Nature*. It is quite as authoritative and its research reports quite as technical, but it usually also contains a number of general scientific articles which are somewhat more accessible to most readers.

Scientific American (75 Varick Street, New York, NY 10013) is published monthly and is therefore not usually as timely as the three weekly magazines, but what it lacks in promptness it makes up in the full and detailed treatment it gives to each subject.

Other scientifically oriented magazines which were consulted on occasion include *Discover, Science News, The Sciences,* as well as a number of journals, conference reports, etc.

Resources

If you intend to take an active part in environmental measures, whether as an individual or as a member of a local organization, you will probably want all the help you can get. Such help is available. Most of the resources available are environmental membership organizations, but there is also a non-membership service which can provide much assistance. That is:

Center for Health, Environment & Justice. (Lois Marie Gibbs, Executive Director, 7139 Shreve Road, Falls Church, VA 22046) publishes many books and pamphlets on every aspect of environmental action, from songbooks to self-help works on conducting research on political opponents, advice on legal questions (and how to find a useful lawyer), help in getting testing materials for environmental dangers, etc. Ms. Gibbs, who founded the Center, became an environmental activist for the best of reasons: when her two children became ill, she discovered that the community she lived in had been built over a toxic waste dump left by the Hooker Chemical Company. It was the famous Love Canal. She has been leading environmental battles ever since.

Because of the rapidly rising ecological interest new national (or even international) environmental organizations are popping up all over. The ones listed below have been around for some time. Almost all of them publish a periodical of some sort, and many of these were consulted in the preparation of this book.

Cousteau Society (P.O. Box 506, Etna, NH 03750) publishes two magazines for its members, *Calypso Log* and *Dolphin Log*. Although its primary concern is with the preservation of "Our Water Planet" the magazines contain material on many other subjects, including population problems, the Antarctic, etc.

Environmental Defense Fund (257 Park Avenue South, New York, NY 10010) is active in legislative and judicial activities on behalf of the environment. It does not publish a magazine for its members but provides them with an environmental calendar.

Greenpeace (702 H Street NW, Suite 300, Washington, DC 20001) is the famous organization which sends out volunteers in rubber boats to get in the way of whalers and ships dumping toxic wastes in the ocean, and whose vessel, the *Rainbow Warrior*, was bombed and sunk in harbor at Auckland, New Zealand, as it was preparing to contest French resumption of nuclear tests in the Pacific Ocean. That is not all Greenpeace does, though. It monitors environmental hazards all around the world and reports them through its magazine, *Greenpeace* (which is distributed to members), and it has available a number of videotapes and study guides which it makes available to interested groups. (Contact Karen Hirsch at the address above.)

National Audubon Society (225 Varick Street, 7th Fl., New York, NY 10014) concentrates primarily on threats to woodlands and wildlife. Its membership is large and influential, though more staid and conservative than comparable groups in the field of conservation. It publishes a bi-monthly magazine, *Audubon*.

National Parks Conservation Association (777 6th Street NW, Suite 700, Washington, DC 20001) is primarily concerned with protecting America's national parks. It publishes *National Parks* for its members.

Natural Resources Defense Council (40 West 20th Street, New York, NY 10011) specializes in legal actions to preserve the environment. It publishes a magazine, *The Amicus Journal*, and a newsletter, *NRDC Hotline*, for its members.

Sierra Club (2101 Webster Street, Suite 1300, Oakland, CA 94612) began as a largely Californian organization with the particular objective of preserving West Coast forests, but has expanded to include members in all 50 states and all Canadian provinces, as well as US possessions and countries around the world. At the same time it has broadened its concerns to include most environmental issues. Its publication, *Sierra*, is distributed to members.

The Wilderness Society (1615 M Street NW, Washington, DC 20036) is also particularly concerned with preservation of national forests and parks. It publishes a magazine for members, *Wilderness*.

Worldwatch Institute (1400 16th Street NW, Suite 430, Washington, DC 20036) devotes itself to researching and reporting on the state of the environment and the world in general. Each year it publishes the volume *State of the World* (distributed to its members), which is the single best annual source of data on new developments in environmental matters.

The following organizations are not basically environmental but can be of considerable value in ways related to efforts to clean up the world—and, in particular, American political and governmental institutions.

Common Cause (805 15th Street NW, 8th Fl., Washington, DC 20005) has as its main purpose the effort to end political abuses and to reform the American political process so that it meets the needs and reflects the interests of all citizens, regardless of party or station in life. Non-partisan and without a political agenda of its own, it has been influential in supporting legislation to clean up campaign financing, ending abuses of power by well-placed legislators and exposing corruption at all levels of government—the precise steps without which no effective environmental legislation can be passed over the opposition of lobbyists and special interests. It is a national organization which also maintains state and local groups.

Center for Defense Information at POGO (1100 G Street NW, Suite 500, Washington, DC 20005) is the best available authoritative source of current information on American military plans and activities which is not directly tied to (and influenced by) the Pentagon and its suppliers. The Center's staff is largely composed of retired U.S. military officers who are experts in the subjects covered. It publishes a small newsletter and prepares videotapes for use on local television stations, which are available to groups.

Federation of American Scientists (1725 DeSales Street NW, Suite 600, Washington, DC 20036) was founded shortly after World War II by a small group of scientists who had been associated with the Manhattan Project and related research projects and were concerned to increase public awareness of the implications of nuclear developments. Its publication, *The Bulletin of the Atomic Scientists*, continues to report on developments in nuclear armaments and disarmament proposals, and also covers such topics as nuclear waste disposal problems, etc.

APPENDIX 2
Constitutions, By-Laws, and Other Parlor Games

You only really *need* a constitution for your club if you incorporate, when some such document will have to be part of your legal papers. (Incorporation is not strictly necessary itself, but it's a good idea if only to save you some financial trouble. If you have a picnic and some kid breaks his leg his parents may sue the organization, or some polluter may sue you for libel; if unincorporated they may be able to collect from the private assets of every member.)

You may *want* a constitution just for the sake of being orderly. Consider, though, the advantages and disadvantages.

What a constitution does is set up the rules by which you play the game of running a club. Most rules, however, are prohibitions; the more rules you have, the more things you may find you can't do. A club without a constitution remains wholly flexible and can decide from day to day what it wants to do next.

Now look at it from the other side: the prohibitions in a constitution may keep you from acting too hastily in a way that you might later regret; or it may keep a small number of people who happen to show up at a meeting from committing the entire membership to something they don't like.

The British get along fine with a constitution that doesn't

exist anywhere in spelled-out form. The Americans, on the other hand, get along equally fine with a detailed written constitution. So either system works.

In any case, if you do decide to have a constitution, you should make every effort to keep it as simple and clear as possible. One rudimentary sample constitution could be written as follows.

Article I: Name

The name of this organization shall be _____.

Fill in whatever name you like, and try not to change it. Remember that you develop property rights in a name. It is what the media call you in your publicity and it is what prospective members know you by. If you change it you depreciate the value of your property.

Article II: Purpose

The purpose of this organization shall be to unite persons who are concerned with the damage to our environment and to provide a means to work toward the control or remedy of that damage.

(If yours is an overtly *political* club affiliated to a party, you might want to replace "persons" with "Republicans" or "Democrats.")

The trick in stating your purpose is to be as general as you can without being wholly incoherent. It is a difficult trick. You are likely to waste more time on this paragraph than on any other single act of your club in its early stages. Possibly the best thing to do is to get each member to write out a statement of purpose and vote on them.

APPENDIX 2

Article III: Membership

Membership in this organization is open to all persons (above the age of ____?) (who are Democrats/Republicans?) (who reside in the area of _____?) *upon application and payment of the first year's dues* (and approval by vote of the membership committee?).

The parenthetical phrases are modifying clauses you may or may not want included for the purpose of restricting potential membership. In general, you don't want to restrict membership very much. It's hard enough to get people to join in the first place.

Article IV: Dues

Annual dues shall be charged (in the amount of X dollars per year?) (at such rate as the vote of the membership shall decide for each fiscal year?) *and no person shall be considered a member in good standing, or be entitled to vote or hold office, unless he has paid the dues for the current year.*

You don't *have* to have dues, but you probably want them, not only to put something in the club treasury but so you can tell who is really a member. You might want to make a special lower rate for students, seniors, etc.

Article V: Officers

The officers of this organization shall be a president, a vice-president, a secretary and a treasurer. There shall also be five trustees, to be elected at the beginning of each (calendar? fiscal?) *year to serve until the first meeting of the following year. These officers and trustees shall comprise the executive committee of the organization, which executive committee shall be empowered to transact necessary business between membership meetings of the organization.*

You might want to spell out the duties of each officer. You might want to have two secretaries, one to keep minutes, one to handle correspondence and mailings. You might want to specify that no more than two officers can come from a single election district (or municipality, or block) of your area.

A good thing to do with trustees is to give each of them a job as liaison with one of your working committees. This keeps them from getting hog-fat and lazy.

Article VI: Committees

The following standing committees shall exist, with the duties and powers appropriate to their functions:

1. A membership committee.
2. A publicity committee.
3. A legislative liaison committee.
4. A house committee.
5. A finance committee.

A chair shall be appointed by the president for each standing committee, to serve until the next election meeting of the organization. The president may also appoint such additional temporary committees as the executive committee or a meeting of the membership shall direct.

(Maybe the president also appoints all the members of each committee—or its chair does—or any combination thereof. Try to get all members, new ones in particular, on a committee, particularly if the club gets large enough to seem impersonal to a newcomer.)

Article VII: Elections

The first meeting in each calendar year (or some other date?) *shall be designated a general membership meeting, and its first order of new business shall be the election of officers for the following year. All voting shall be by secret written ballot, and no proxies or absentee ballots shall be counted.*

The secret ballot can be waived on the request of the chair for unanimous consent when there is no contest, but it is a mighty aid to democracy when a contest does appear.

The question of whether or not to allow proxies is vexed. They're good if some desirable members simply can't be present and you don't want to exclude them; but it's better not to allow the use of proxies if there is any kind of friction in the club, because some dissident member may turn up with enough to cause trouble. If worst comes to worst, flip a coin.

Article VIII: Procedure

Robert's Rules of Order shall determine all procedural questions not otherwise determined in these articles.

You might want to go out and buy a copy and then specify that edition as the particular one you mean; the editions do vary slightly.

Article IX: Amendments

This constitution may be amended by simple motion of any member in good standing, passed by an affirmative majority vote of all members present at two consecutive membership meetings of the organization.

The idea is to make changing the rules slow and difficult, but not impossible. . . . And that's about all you need.

Do do everything in your power to write and adopt a constitution at the earliest possible moment, while you have relatively few members. The more people are involved, the longer your constitution will be (as members contribute their memories of constitutions from organizations past) and the harder it will be to get consensus.

Remember, it doesn't really matter all that much. All the constitution does is set the rules of the game, and you can play by one set of rules nearly as well as by any other . . . just so you don't keep changing them.

AFTERWORD
One Year Later

Frederik Pohl

The really bad news about the environment hadn't happened at the time when this book was first written, and it hasn't all happened even yet, a year later. To be sure, there certainly are people and animals in the world today who have been blinded with ozone-loss cataracts, and there are some who are dying of skin cancer because of that same destruction of the ozone layer. But the numbers are still so tiny, relatively speaking, that it is almost impossible to identify them accurately because they are lost in the (as yet) still much larger normal incidence of such illnesses. There are weather changes, too; there is continuing destruction of forests and other natural resources; there are more and more species of living things going extinct. All those things are happening now, but the truly menacing, large-scale, even devastating probable consequences of what we are doing to "our angry Earth" are still in the future.

That is not really very comforting. Unfortunately for all of us, what is in the future doesn't *stay* in the future. The future keeps coming closer and closer to us, at the unalterable rate of 24 hours each and every day.

The first edition of *Our Angry Earth* reached the bookstores in November, 1991. Books do not magically appear on the shelves as soon as their authors have finished writing them. It takes time to set type, time to correct proofs, time to print and bind the volumes, time to get them out of the warehouses and into the stores. Working right up to the last possible deadline, it was four months before the publication date—that is, it was in July, 1991—when Isaac Asimov and I had our last chance to update the manuscript.

Now it is bit more than a year later.

So we have a chance to perform an experiment. If things are really going to be as bad as all the data suggest, we are now a year closer to the bad times. So it is worthwhile taking a look at what has happened in that past year to see whether the predictions appear to be working out, or whether somehow all the climatologists and other scientists from whom we drew our conclusions were mistaken and their warnings were without foundation.

That is why this additional section has been written for the book's new edition: to take the evidence of one additional year and use it to try to see, as best we can at this still early date, whether those predictions are on track.

Sadly, this last part can no longer be a collaboration between Isaac Asimov and myself. Isaac is no longer here to make it so.

For more than half a century Isaac was my very good friend. We became friends when we were both barely seventeen years old, two teenaged science-fiction fans and would-be writers who shared a burning, but as yet unfulfilled, desire to get published. We wrote a couple of short stories together in those early days (they appear in the collection *The Early Asimov*), and over the years we worked together now and then in many ways. For a time I was his literary agent, and occasionally I was fortunate enough to be his editor. However, we never collaborated on a writing project again until, toward the end of 1989, we decided to join forces in writing *Our Angry Earth*.

We celebrated that decision with a dinner in Isaac's favorite restaurant, Peacock Alley in the Waldorf-Astoria Hotel in New York City. Our wives, Janet and Betty Anne, were with us, and it was a fine and festive occasion. The only thing to mar it came at the very end when, as Isaac was getting up to leave, he announced that he was beginning to feel a little unwell. He thought, he said, that he might be coming down with a touch of flu.

It wasn't flu, though. It was the recurrence of a defect in the mitral valve of his heart. From the restaurant Isaac went back to his home, and to bed, and shortly thereafter to the hospital. He rallied for a while—more than once, and sometimes for months at a time—but the strain on his heart, added to the cumulative strains on his system that came from all the complications involved, caused relapse after relapse. And at last, in April of 1992, at the age of 72, Isaac died.

Years ago a friend had some calling cards printed up as a gift for Isaac. They said simply:

Isaac Asimov
National Resource

I think the cards were intended as a sort of joke, but they were more than that. They were a quite accurate description

of a valuable, brilliant, well-loved and entirely irreplace-
able man.

Global Warming and the Weather

We have discussed many environmentally destructive processes in this book, but among them there is one which stands
out. It isn't only the largest in scale. It is also the one that
has produced the most prolonged and furious debate. That
one, of course, is the prediction of a general global greenhouse warming brought about by human activities.

It is no surprise that so much contention should arise on
this issue. A great many influential people don't want to
believe in the process at all, because of the imagined consequences. What they fear is that any serious attempts to
forestall the warming by reducing the human production of
greenhouse gases will require major changes in the way we
live our lives and run our industries. (As in fact they will;
though, as we have already seen, they need not be overly
expensive or difficult ones.) Moreover, weather conditions,
temperatures included, change unpredictably from year to
year because of the action of natural processes like the El
Niño-Southern Oscillation ("ENSO" for short) in which the
waters of the south Pacific seem almost to slosh back and
forth over a period of months or years. For these reasons it
is inevitably difficult to try to untangle any clear evidence of
a warming which can be uncontestably blamed on human
activities.

What made a hard task harder still was the eruption of
the Philippine volcano, Mount Pinatubo, on June 16, 1991.
Pinatubo's eruption was very violent. It cast its cloud of sul-

furic acid particles high into the stratosphere. There they became solid particles of sulfur compounds, thus creating a sunscreen which reflected some sunlight back into space which would otherwise have gone to warm the Earth's surface.

To help clarify this question, Ellsworth Dutton of the National Oceanic and Atmospheric Administration tried to untangle Pinatubo's effects in the summer of 1992. When he analyzed global temperature readings for the months of April and May, 1992, he found that the Earth's temperature had fallen by an average of about one degree Fahrenheit worldwide for that period. That global average did not give the whole picture, though. It concealed some large local differences. The drop in the northern hemisphere as a whole amounted to a degree and a half Fahrenheit—as would be expected, since Pinatubo lies in the northern hemisphere and air circulation across the Equator is comparatively slow. Yet France, also firmly in the northern hemisphere, bucked the trend by suffering its *hottest* May in half a century, while in the northeastern United States and eastern Canada it was even colder than the hemispheric average, with the average temperature for the region depressed as much as 8 degrees Fahrenheit below normal in that same month of May.

Although Mount Pinatubo's was the most violent volcanic eruption in many years, it was not unique. Fortunately for climatologists there is enough history of similar eruptions in the past to give some idea of how long its effect may last. Projections based on this historical experience indicate that worldwide there will be a drop of one additional degree in the second year after the eruption, after which the particles that make up the Pinatubo sunscreen should fall out and global temperatures should begin to return to normal—or, more probably, to the higher than normal temperatures associated with the global greenhouse warning.

But—nothing is ever easy in these matters!—there are two other factors which may alter that scenario.

For the first, there are some indications that Pinatubo may (as of September 1992) be on the verge of a second major eruption. More troublesome still, it now appears that one of the human activities that plays a part in bringing about the global greenhouse warming may also have a duplicitous role in masking it as well.

One of the arguments used to cast doubt on the computer models predicting warming is that, although there is no doubt that some warming has been detected, the measured increases are not as great as most of the models predict. Opponents of the warming theory have seized on this to suggest that therefore the models must be wrong.

However, a new study published in *Science* in June 1992 by Joyce E. Penner and other scientists at Lawrence Livermore and the University of Arizona suggests that instead the models are only incomplete. Penner's team looked into the sun-shading effects of smoke from biomass burning, of the sort used to clear tropical forests for agriculture. Although the smoke clouds are patchy and the aerosol particles they contain are short-lived, the Penner team's analysis showed that they did in fact have a total worldwide cooling effect "of an amount comparable to the warming expected from a doubling of CO_2."

Of course, such a neat balance between promoting warming (by adding to the burden of greenhouse gases as the woodlands are burned into carbon dioxide) and disguising it (by preventing some of the Sun's warmth from reaching the surface of the Earth) has to be temporary. Such a tightrope walk could not prevent the warming forever; sooner or later we would run out of forests to burn. And in any case the pre-Pinatubo record remains: most of the ten warmest years in the global historical record occurred in the 1980s, while the year 1990 was the warmest of all; and the first half of 1991 looked as though it would turn out to be warmer still, until Pinatubo blew its top.

Finally, the most influential statement supporting the global-warming theory came from the Intergovernmental Panel on Climate Change, representing the majority of the world's climatologists, which reached its first consensus in 1990.

In January, 1992, the Panel met again, this time in Guangzhou, China. Their purpose in reconvening (like ours in this update) was to try to determine what eighteen months of additional research and observation had done to support or refute their predictions. Their 1992 conclusion: Although recognizing that the evidence was still far from complete, they agreed that the new data "either confirm or do not justify alteration of the major conclusions of the first IPCC assessment."

As we've discussed earlier, some of the anticipated consequences of global warming include melting of the Earth's ice stores, sea-level rising and an increase in erratic and even violent weather. It is worth looking at the record to see to what extent those things are happening.

There is, as it happens, one clear case of sea-level rising reported in the last year. The report comes from the Scripps

Institute in southern California, and it states that in the 42 years since 1950, over a 77,000-square-mile region of ocean off the California coast, the average water temperature has risen about a degree and half Fahrenheit, and the corresponding sea level has risen by between one and one and a half inches. Whether this is due to global warming or to some natural fluctuation is unsure. The best that a proponent of the warming could say is that at least it goes in the right direction to support the prediction, although it is in any case still too early to expect very much change in that regard.

As to melting ice, there is no doubt that the glaciers in the Swiss Alps are continuing to melt. That's why, in 1991–1992, they have retreated so far as to disgorge a number of frozen corpses. These bodies are the remains of human beings who were unlucky enough to be trapped in the glaciers at some time in the past—sometimes in the very far past, since the retrieved bodies include one Bronze Age hunter who had lain frozen for nearly four thousand years.

The record for the rest of the world's ice is less convincing, in particular in regard to the polar caps. One bit of evidence, however, suggests that an unusual number of icebergs were calved from the Antarctic ice cap in the past year: For the first time in living memory, numbers of huge bergs, as much as 125 feet in height, appeared ninety miles off the coast of Uruguay, only 35 degrees south of the Equator, in June, 1992. (Again this, while not proof, is at least what would have been expected if the warming were real.)

How about the predicted erratic and violent weather?

Before we answer that, we should repeat the caution: Anecdotal evidence, which is to say reports of odd events in various places, is not usually enough to confirm or disprove a theory. But when it comes to weather that is often all we have. We can't do controlled experiments, because we don't have another planet available to use for a control. We can count up the total number of days or inches of rain in a year that fell on some reporting station, but knowing the amount of precipitation in one place does not necessarily tell you anything about the precipitation in some other place only a few dozen miles away. And, anyway, what is "unusual" about the weather is a judgment call.

Still, the anecdotes do show some troublesome weather patterns. Droughts continued in Africa and Europe. The drought in southern Africa looked to be spreading north, and famines began in Zimbabwe. Zimbabwe was not used to famine; it usually raised enough food to export a surplus—but as that surplus had been exported for foreign exchange it had no reserve to meet the new threat, and observers noted that the hair of many Zimbabwean children was beginning to turn red, a symptom of the deficiency disease, kwashiorkor. North Africa was in its sixth year of famine. Much of the suffering here was caused by fighting in the area's interminable civil wars, preventing outside help from getting to needy civilians, but 2,000 people a day were starving to death in affected areas.

In England 41 consecutive dry months were interrupted by one wet one in April, 1992, but in May the drought resumed. It affected only the urban southeast section of the country, including London, while rain was adequate in the lightly populated northwest, but the dry area was suffering its longest dry spell since 1745. The River Darent in Kent was on the verge of drying up completely; the Ver in Bedfordshire had already done so in part, turning into cracked, dry, black mud at St. Albans (where the Romans used to anchor their galleys), and the river's populations of kingfishers, trout and watercress were gone. As of the summer of 1992 England was contemplating such emergency measures as a National Rivers Authority project to line riverbottoms with clay and synthetic cloth to cut down seepage, while the Thames Water Authority, which supplies London, announced plans to pump treated effluent from its sewage plants back upstream to be put back into the river, in order to keep its flow up to a tolerable minimum.

France was in its fourth year of drought; French rivers shrank, too, and the beds of rivers like the Loire at Nantes also became large expanses of cracked and dry mud. In Spain half the rice crop was lost, and 17% of the area of the country was now officially classified as desert.

Droughts in one part of the world were offset by unprecedented floods in others, for example in Sri Lanka. Colombo, the capital of the island nation, is a tropical city which gets most of its annual rainfall—about forty-odd inches—in the monsoon season. It began to rain hard in Colombo on the evening of June 4th, 1992, and when the deluge ended eighteen hours later rivers had overflowed their banks, homes were washed away and the Sri Lankan Parliament buildings were flooded with several feet of water. The rainfall for that short period amounted to 22 inches—half a monsoon season's worth in less than one day. They called it a "once in a thousand years" rain, but in fact there was no record of such a drenching downpour ever in the long history of Sri Lanka. Then, in the late summer of 1992, Pakistan, too, suffered devastating floods, with more than 2,000 villages and small towns inundated, most of the cotton and rice crops lost and the giant dams on the Indus River threatened with collapse; the death toll was uncertain but as many as 5,000 lives may have been lost.

America had unusual weather of its own. In the American midwest, for instance, the weather of 1991–1992 was exceptionally mild. Temperatures were warmer than usual, and heavy snows almost nonexistent—with two curious exceptions. Oddly, two of the worst snows in the region occurred in the wrong seasons. One was in the city of Minneapolis, which in early November, 1991, set a new city record for the heaviest one-day snowfall in its history, sixteen inches in twenty-four hours—but that blizzard actually happened six weeks before winter officially began. While the Chicago area's heaviest snowfall of that winter actually smote the area on what the calendar said was the first full day of spring.

The spring was equally unusual. We've already mentioned the unusually cool May and June in the northeastern United States; but there was more. The month of May, 1992, for instance, was the wettest ever seen in El Paso, Texas, as moist monsoons came up from the Gulf of California, while it was the *dryest* May ever—averaging only about a tenth of normal precipitation—in a wide slash that ran across the country from the Pacific Northwest through the Chicago area and much of the rest of the midwest and on down to Miami Beach.

July was another exceptional weather month. On July 2 a series of funnel clouds and tornado touchdowns began in Iowa and Illinois, moving eastward with the storm front. Large cities are usually unlikely to suffer tornado-type winds, probably because of the "heat island" effect, but in Chicago a near tornado raced through the city's annual "Taste of Chi-cago" food festival along the lake-front, collapsing refresh-ment tents and capsizing a score of sailboats in a children's regatta; fortunately no lives were lost there, though elsewhere one woman was electrocuted by a downed power line and a new cinderblock wall toppled on a crew of workmen. Yet July of 1992 was also one of the coldest Julys on record in eight states in an arc from Nebraska and Iowa east to Massachu-setts and New York, while Florida and the Carolinas were unusually warm. (Nationwide, the normal July average tem-perature is 74°F; in 1992 it was only 71.7°, the third coldest July on record in spite of those few exceptionally warm areas.)

Do all these individual weather oddities add up to any meaningful meteorological pattern?

The answer to that question is still a judgment call. Those who are convinced by the warnings of global warmup may detect a trend; those who aren't still have the privilege of dis-missing them as the sort of meaningless aberrations that occur in many years.

Still, we haven't yet told the whole story. What about the prediction of bigger and worse hurricanes?

That prediction looked doubtful in the early stages of the 1992 hurricane season, because it was the sparsest in mem-ory. Not a single Atlantic hurricane appeared until mid-August.

Perhaps it was the sunscreen from Mount Pinatubo that kept the early storms from developing. However, when that first hurricane of the season came along it was a monster.

Hurricane Andrew did more damage than any other sin-gle natural disaster in American history. It wasn't the strongest hurricane that ever hit the United States; that strongest category is called a "Class 5 hurricane" with sustained winds over 160 miles an hour, and for most of its life Andrew's wind velocities hovered just below that. (Though it did reach Class 5 now and then, and the top velocity of Andrew's winds near Homestead, Florida, where it did the worst of its damage, is not known. The storm blew away the instruments at nearby Biscayne National Park after the winds reached 167 mph.) Yet the dollar cost of Andrew was estimated at $20 billion, more than triple the losses from the second worst storm, Hurricane Hugo in 1989.

The weather service did itself proud on Hurricane Andrew. It forecast its path with great accuracy. Yet, in spite of timely and reliable warnings, some 20 lives were lost. Andrew completely destroyed the Florida communities of Homestead and Florida City, as well as causing major destruction in the cities of Miami and Miami Beach. And, of course, it also wrecked large areas in the Bahamas before hitting Florida, and in coastal Louisiana afterward, though the dollar losses were lesser simply because there was less in those areas to destroy.

In Florida more than a million people were left without water, because of sewage contamination in the mains, and even more without power or phones. Rail tracks were twisted out of the ground, putting the mass transit system out of operation. Andrew even destroyed two primate research centers in the Miami-Homestead area and set nearly two thousand monkeys free to roam the storm-devastated area and add to the confusion. Many of the monkeys were part of a National Institutes of Health program for developing a colony of monkeys free of retrovirus to be used in AIDS research. Local people, hearing distorted versions of this, suspected the monkeys were carrying AIDS, so they took their shotguns off the wall and shot as many of them as they could find.

Although Andrew was the first real Atlantic hurricane in the 1992 season, the Pacific was more active. There is usually less damage from a Pacific hurricane than from an Atlantic one, simply because there is so little land mass in most of that great ocean. Still, shortly after Andrew did its work in Florida and Louisiana, Hurricane Iniki clipped the western edges of the state of Hawaii, killing three and doing another billion dollars' worth of damage. The island of Kauai suffered the most, where Iniki was the worst such storm in a century, destroying pineapple and macadamia-nut plantations and wrecking some of the island's ornate tourist hotels. It was, Senator Akaka of Hawaii told his Senate colleagues while pleading for disaster relief, "a scene of unimaginable devastation."

That was not the whole Pacific toll for 1992. There were dozens of such storms in that ocean, doing considerable damage in other countries, and even one more American possession had suffered when a hurricane named Typhoon Omar hit Guam.

(Yes, "Typhoon" Omar was a hurricane. That's what a typhoon is; it is only a matter of convention that, if a great tropical storm of that kind happens to occur east of the International Date Line, it's called a hurricane, while if west of the Line it is a typhoon. The reason for having two names for the identical phenomenon is simply that European languages had no word for these storms—had no need to have one, since they almost never happened in Europe. The reason for *that* is geography. Hurricanes begin as mild tropical depressions in the eastern reaches of large oceans. Natural circulation pushes them westward. As they move in that direction they gain strength from the warm sea waters, attaining devastating force by the time they impact whatever land lies on the western edge of the sea. Since Europe doesn't happen to have an ocean to its east, it was spared these repeated and highly destructive seasonal storms, and when European colonists first encountered them in America and the Pacific they took the names for them from the languages of local people.)

Hurricanes, of course, are not totally evil, even from the parochial point of view of the human race. They do a great deal of damage, but they also are an important source of rainfall for the interiors of continents like North America, where the transport of moisture from ocean to land in the hurricane season provides a substantial fraction of the year's precipitation; without hurricanes there would be more droughts, and far worse ones. And they have other virtues as well.

Indeed, if things had gone a little differently Hurricane Andrew could have performed a great service for south Florida, where Florida Bay has been stagnating for many years. Deprived of its fresh-water flow from the Everglades because of diversions to meet the needs of the area's rapidly growing population, Florida Bay has become heavily polluted with sediment and nutrient waste. Its salt content has increased, damaging marine habitats, and algae blooms have carpeted twenty-square-mile areas of the bay.

A strong, sustained hurricane could have torn the bloom away and flushed the bay out. But, violent though Hurricane Andrew was, it went through too fast; it did not carry quite enough rain to dilute the bay waters; and its winds were too localized to destroy the algae bloom. Florida Bay is still stagnant and dying.

The Ozone Layer and Its Consequences

If the evidence on the global warming is still basically inconclusive (though apparently trending in the expected direction), the story on other environmental disasters is far more definite: for example, what is happening to our ozone layer.

There the news is all bad—even unexpectedly bad, in ways that no one had anticipated. Although depletion of the ozone layer at the poles, particularly over the Antarctic, had been well documented at the time the bulk of this book was written, it was still then generally supposed that most inhabited portions of the Earth had some time to prepare for the consequences of ozone loss. That confidence was shattered when, in the fall of 1991, a new NASA study revealed that ozone had been depleted over a broad band of the Earth's surface ranging from 40 to 50 degrees of *North* latitude.

That discovery was wholly unpredicted. The ozone-depleted band covered many of the most heavily populated regions of the world. In North America it included such cities as New York, Boston, Toronto, Detroit and Chicago, all the way across the continent to Portland, Oregon.

Although this temperate-zone ozone loss was not nearly as great as the one that had been discovered over Antarctica, it represented a loss ranging from 10% to as much as 40% of the ozone shield in some seasons. The cause is still debated, with the prevailing opinion suggesting that here, too, the sulfuric acid emissions from Mount Pinatubo have played a part, probably helping to release the ozone-destroying chlorine compounds from the burden of CFCs we have manufactured and discharged into the atmosphere.

Perhaps even more surprising, a later study showed that the tropics themselves were not immune to the growing areas of ozone loss. Even near the Equator ozone was depleted as much as 20% at some times.

Nor has the Antarctic ozone hole healed itself. On the contrary. In February, 1992, Dr. Susan Solomon of the National Oceanic and Atmospheric Administration reported to the annual convention of the American Association for the Advancement of Science that in the past season ozone over the Antarctic was even more depleted than expected. 83% of the ozone at altitudes from 7 to 12 miles above McMurdo Station was gone. At some elevations the loss rose to 93%.

Of course, it is not the depletion of ozone itself that threatens our well-being, it is the consequences that will flow from the loss of that protective shield against the harmful ultraviolet-B radiation from the Sun—particularly what it will mean in terms of increased skin cancers.

Here a number of new studies have given us a clearer picture of what may be in store for us. To summarize them:

There are three basic types of skin cancer that are affected by ultraviolet-B from the Sun: basal-cell carcinoma, squamous-cell carcinoma and melanoma. They are not all affected in the same way, and they do not all present the same threats to health. The two carcinomas are usually operable and, with appropriate surgical treatment, are seldom fatal, although they may be disfiguring. Melanoma, on the other hand, often kills.

As to, for instance, the squamous-cell carcinomas, it is not exactly true to say that ultraviolet-B radiation *causes* them. Such cancers can arise for other reasons, and perhaps spontaneously for no identifiable reason at all. Such ordinary carcinomas do not usually grow to a harmful size, however, because the body has a defense against them, which is controlled by the tumor-suppressing gene called *p53*. The dangerous role played by UV-B is that it causes the *p53* gene to mutate into a biologically useless form, and then it can no longer prevent the spread of the cancer.

With the new studies, it is now possible to put some tentative numbers into these general principles. Roughly speaking, a 1% loss of ozone allows approximately 2% more UV-B to reach the surface of the Earth below. That 2% increase in UV-B is expected to produce some 8% more squamous-cell carcinomas, and about 4% more basal-cell carcinomas.

We do not yet have that sort of detailed understanding of the melanomas. Scientists are still uncertain as to whether the ordinary "harmless" UV-A may not be just as important as the new UV-B in producing them, because the exact mechanisms producing them are complex and still poorly known. For one thing, it seems to be the case that adults who suffered more than one or two bad, blistering sunburns in childhood—long before the present thinning of the ozone layer and the consequent increases in UV-B radiation—are at greater risk for developing melanomas than those who did not.

Ultraviolet-B, however, definitely does increase the number of these cancers. Accordingly, on balance, the scientists at the Skin Cancer Federation predict that for every 1% loss of ozone the number of melanomas, including lethal ones, will also increase by somewhere between 1 and 2%.

Finally in this segment of the ozone story, we are reminded that everything in our environment is interconnected: Increasing ozone loss will have a direct effect in accelerating the global greenhouse warming.

In *The Bulletin of the Atomic Scientists* (June, 1992), Jeremy Leggett points out that, actually, more carbon dioxide is taken up from the atmosphere by the tiny marine organisms called phytoplankton than by all the world's land vegetation combined. They account for an uptake of something over 100 billion tons of carbon a year, though we land-living human beings seldom think of the oceans as so heavily vegetated.

Unfortunately, the phytoplankton are particularly sensitive to UV-B. Worse, they have no way of avoiding it; they float passively near the surface of the sea and have no shelter.

How severe will the damage be? Clear quantitative evidence is lacking, but Leggett quotes a new U.N. Environment Program report which suggests that "a hypothetical loss of 10 per cent of the marine phytoplankton would reduce the oceanic annual uptake of carbon dioxide by about 5 billion tons—an amount equal to the annual emissions of carbon dioxide from fossil fuel consumption."

The Rest of the Record

Not unexpectedly, most of what we have to talk about for the past environmental year is bad news. Still, there is one small bright spot—well, one that is a little brighter than expected, anyway. It turned out that the long-range consequences of the Gulf War were not quite as devastating an ecological disaster as they might have been. The effects of the smoke from the Kuwaiti oil wells Saddam Hussein set ablaze as he retreated could have been a lot worse.

The world was spared some of that anticipated smoke damage because the burning was over with relatively quickly. The highly efficient teams of oil-well firemen from Texas and elsewhere suceeded in putting the flames of those 749 torched wells out a lot faster than anyone had guessed. Most experts had predicted that the task would take the better part of a year; in the event, the last fire was extinguished in November, 1991. (Actually that was the second time that particular oil-well fire was extinguished. The experts had routinely put it out some time earlier. Then, at the request of the Kuwaiti authorities, it was re-ignited so that one of their high officials could have his picture taken in the act of extinguishing the "last" one himself.)

The more important reason the damage was not worse was that the devastating smoke pall simply never rose high enough.

If it had risen to a height of nine miles, instead of only about half that elevation, it would have reached the stratosphere, where the high-level winds would have carried the pall of dense smoke far away. Then, instead of merely turning day into night in Kuwait City, it would have shaded the Sun over much of southern Asia. That might well have caused a "mini–nuclear winter," as Carl Sagan put it, and places as distant as India might have had to pay a high price in crop losses and hunger for Saddam's petulance.

Why were the projections wrong? Why did the smoke stop short?

Since no such large-scale oil fires had ever happened before, the scientists were working in the dark. On the basis of what data they did have they assumed the smoke cloud would be "self-lofting"—meaning that it would rise as it was warmed by the Sun—but it wasn't. They expected the cloud would remain airborne for a long time, but as it failed to reach the expected heights, it didn't. They supposed the combustion from the wrecked wells would be inefficient and incomplete, producing very sooty emissions like those from a badly regulated furnace, but in fact the burning was more thorough than predicted, and thus the cloud's sun-shadowing content of large unburned particles was less. If the world is ever unfortunate enough to have another such incident the scientists will be able to make better predictions, because then they will have the knowledge gained from Kuwait to guide them. But they didn't have that knowledge at the time.

They were, however, on target about the local environmental damage. In Kuwait itself the smoke pall cut off three-quarters of the warmth from the Sun, and droplets of oil that condensed out of it fell to coat the ground with a layer of tar. Fisheries in the Gulf were badly damaged by smoke fallout as well as by the crude oil Saddam discharged into the sea. And, of course, none of that takes into account the vast damage to the environment that was caused by the particularly violent war that was fought over that territory.

That is about the last of the environmental news that can be called even qualifiedly "good."

In most other respects, a year's passing has left most other environmental affairs worse than ever, or at best little changed. It still takes more than 600 gallons out of our diminishing potable water supplies to produce the meat in a single quarter-pounder hamburger. We're still burying some of our best agricultural land under road-building (by mid-1992 nearly forty thousand square miles of American soil had been paved over for streets and highways—an area the size of an average state). We still continue to drive species and subspecies to extinction (1992 was the last year for the Japanese crested ibis, and at the Idaho end of the Snake River, where thousands of sockeye salmon used to swim 900 miles upstream to spawn, only four fish made it alive to the spawning grounds in 1991.) Even orbital space is still littered with dangerous amounts of our discarded trash: in September, 1991, the Space Shuttle *Discovery* had to take evasive action to avoid collision with an abandoned, but still orbiting, Soviet rocket stage.

Since so little has changed, in spite of all the environmentally pious pledges of the world's leaders, there is not much reason to say again all the things we've already said on these subjects. But we might quickly recheck a few of the major disaster areas:

Forest clear-cutting. The rate of deforestation in the tropics and in the American northwest has not slowed, while the world's hunger for fresh supplies of wood has continued to grow. Now some vast new tracts are under the axe.

The largest remaining forests in the world are in Siberia. There are more than two million square miles of them, an area as big as the continental United States, and over half of that huge area has never been logged. Until now.

In Russia's desperation for foreign exchange to help its economic misery it is selling almost everything that can find a market abroad. Even parts of its once-sacred weapons systems have been offered for dollars from its former enemies, and inevitably its forests are also for sale. Some have already been sold and are gone. The South Korean Hyundai Corporation has already begun massive clear-cutting in the prize spruce, fir, larch, pine, oak, ash and elm forests along the Pacific coast, and is negotiating for even larger claims inland. So are many other foreign corporations, including America's Weyerhaeuser. And, in addition to what logging still more trees will do to the carbon-dioxide levels in the atmosphere, there is another acute environmental hazard in this case. Much of Siberia's prime forested areas stand on permafrost, which is particuarly vulnerable to clear-cutting. Past experience indicates that about half of such areas do not regenerate after logging. They turn into swamps.

Elsewhere, clearcutting the Philippine forests (where nearly 60% of the woodlands had disappeared since 1930) has left the islands' exposed mountain soils unstable, so that when Typhoon Thelma hit the island in 1991 many of the 6,000 dead were killed in flash floods and mud slides from the unprotected slopes. Meanwhile, Thailand's lumber interests, forbidden to continue commercial logging in their own country after landslides killed 430 people in 1989, simply moved across the borders to repeat the process in other nearby countries.

Extractive damage. Open-pit mining continued to destroy fertile areas around the world. One particularly destructive example became public in June, 1992, when the now independent island nation of Nauru brought damage suits against its foreign former administrators in the International Court of Justice in The Hague.

Nauru is a dot in the Pacific Ocean no more than ten miles across, close to the Equator. It originally possessed one tremendous natural resource—deep layers of phosphate, a commodity in worldwide demand as fertilizer, that covered most of the island—but now the phosphate has been mined and shipped away, and ninety years of strip-mining have left 90% of the island bare coral, where nothing can be made to grow. Only a thin fringe along the coasts of Nauru is still inhabitable, and as Nauru has no other industry the future of the nation is in doubt. According to David Vincent, an Australian economist, "The best solution may be to abandon the island."

Pollution. Just as before, we continue to discharge toxic and radioactive wastes into our waters, air and soil, while now the bills for previous acts are coming due. For example, the long-range costs of the widespread radioactive pollution that resulted from Chernobyl: In April, 1992, the Ukrainian health ministry announced that thyroid cancer in Ukraine and the nearby republic of Belorus is now three and a half times as prevalent as before the Chernobyl accident in April, 1986. And on March 24, 1992, the Russian RBMK nuclear power-plant reactor at Sosnovy Bor, near St. Petersburg, had a containment accident that released a "serious" amount of radioactive material into the environment. The Sosnovy Bor reactor is of the same type as the one that blew up at Chernobyl in April, 1986 (and also the same as the second Chernobyl reactor which suffered a dangerous fire in 1990). However, unlike the case of the 1986 Chernobyl reactor, the safety systems at Sosnovy Bor had not been deliberately turned off by the operators, and so the reactor automatically shut itself down without going·on to an actual large-scale explosion.

News of an unexpected new problem of dangerous radioactive waste emerged at a United States Senate hearing in July, 1992, when American CIA and Russian diplomatic sources revealed that the Soviet Union had secretly used the Arctic Ocean as a nuclear dumping ground. In all, 2,100 containers of nuclear waste, dumped reactors from 12 nuclear submarines and three nuclear icebreakers (as well as one experimental nuclear sub, the K-27, scuttled whole) and the debris from 130 nuclear detonations were dumped into the sea near Novaya Zemlya. The dumped waste by Russian government estimates totals over than a billion curies of radioactive materials—more than ten times the amount released by the Chernobyl explosion. (England and other northern nations worried about some of that leaked radioactivity drifting toward their shores . . . but Hugh Livingston, of Woods Hole, after studying such flows in northern waters, reported that he found no movement of radioactive water coming toward the British Isles. On the contrary, what he found was a radioactive flow moving in the other direction, probably from waste water emanating from England's Sellafield nuclear installation.)

Military pollution, radioactive and otherwise, is of course not limited to the former Soviets. According to a 1992 statement from the Science for Peace Institute in Toronto, "10 to 30 per cent of all global environmental degradation can be attributed to military activities." Here again, according to the International Physicians for the Prevention of Nuclear War, the bills are just now coming due, for they estimated in 1992 that nuclear bomb testing will cause an additional 2,400,000 deaths from cancer.

In non-nuclear waste, the military leads the pack. The American defense services alone burn about 200 million barrels of petroleum products a year, more than twenty times the amount consumed by all the nation's public transportation systems combined. A single F-16 aircraft burns in an hour what an average American motorist burns in a year, and a 1992 study by the Energy Technology Support Unit at Harwell, England, shows aircraft of all kinds are major producers of that other form of pollution, the carbon dioxide that contributes to global warming. According to their report, air passenger travel produces more global-warming carbon dioxide than trains or even private cars. Per seat mile, air transport produces two pounds of carbon dioxide, while even private cars produce only about four ounces and electric trains little more than an ounce and a half.

Air pollution, largely from transportation, produced choking smog as usual in many of the world's cities, with Athens, Greece, having a particularly bad year. Previous problems had led the Athens city government to impose emergency measures—like banning private cars from the center of the city—whenever the nitrogen dioxide content of the air rose over 500 micrograms. At the end of September, 1991, when temperatures neared 100°F and there was no wind, the levels reached 561 micrograms, the highest ever recorded. The chief culprits were the exhausts from private cars, and especially from the city's fleet of Hungarian-made buses. In that September, 1991, event two hundred people were hospitalized, many fainted at work, and local radio stations were calling the city a big "gas chamber."

And it may be that none of this is the most worrisome pollution news of the year. In September, 1992, a study at the University of Copenhagen, Denmark, reported that sperm counts in healthy men had dropped nearly fifty percent since World War II. The new research supported previous reports which indicated a continuing decline in male fertility, and in both cases the cause was thought to be the long-range health effects of pollution.

Is anything at all being done about all of this?

Well, yes. In fact, more and more of the world's people are beginning to become environmentally conscientious, and putting their principles into practice. Many nations, and some lumber companies, have begun reforestation programs, and in some cases private institutions have taken up the work. In Kenya, the Green Belt movement hires poor women to plant seedlings. For every one that survives, the woman is paid four cents U.S. But reforestation of all kinds is coming under increasing scientific questioning; there is evidence that trees need the spongy water-holding capabilities of old dead logs to maintain moisture, and may require the sort of lichens that grow only on mature trees to fix nitrogen and permit growth. The most environmentally conscious among lumbering enterprises are now experimenting with selective logging, leaving many old trees standing and most of the floor detritus in place.

Technology is beginning to produce new environmentally helpful designs. In California, which has mandated the introduction of zero-emission new cars before the end of the century, many new models are under development—including the "series-hybrid" combination gasoline-electric car, which does burn fossil fuel but because of its design needs only about a third the horsepower of the current gas-guzzlers. Since one of the great markets for forest products is paper, alternative crops are beginning to be grown: the plant called kenaf, which actually produces stronger and whiter fibers than wood; even an old familiar (but surprising) one in the form of hemp, which yields four times as much fiber per acre as woodlots. Fisheries and bird habitats which have approached or even reached extinction are being restocked from hatcheries; and many communities throughout the world have begun mandating separate collections for recyclable paper, glass, plastic, aluminum and other forms of waste.

All that is certainly very good . . . but not, really, good enough. It is far better, for instance, to try to preserve a natural salmon run than to try to restock it. "Tame" hatchery fish do not survive as well as wild-born ones; their death rate is ten times the natural one. The hemp which would make such good paper is prohibited by law: its flowering top is marijuana, and it is a felony to grow it anywhere in America. And the recycling movement, especially as regards paper, is failing because paper mills still refuse to buy the waste.

So even most of the good news is tarnished by failures. That's a disappointment . . . especially since 1992 was supposed to be the year when all the governments of the world were going to get together to take real action about our environmental problems at the Rio de Janeiro conference.

The Rio de Janeiro Conference

The idea of the 1992 Rio conference was to get the countries of the world together—all of them—and sit their leaders down and knock their heads together until they came up with global plans to deal with the global environmental disasters.

It was a grand plan, and not a cheap one. 35,000 people came to Rio for the conference—for the two concurrent conferences, to be accurate, since many of them were individuals and small groups that ran their own unofficial counter-conference to demand action from the formal one. Getting all those world leaders and private citizens to Rio de Janeiro, and housing at least the official delegates in appropriate luxury while they deliberated, cost close to $500 million, one attendee estimated, with the host country of Brazil putting up nearly half a billion more—money that it could not well afford to spend, since Brazil's treasury has been running in the red for decades; in some recent periods its inflation rate was running at six percent—per *day*.

There were some geographical objections to choosing Brazil as the venue for the conference, for it is not an easy place to get to from most of the rest of the world. Still, it was an appropriate choice for this conference. Brazil is a uniquely First World/Third World melange. It has all the environmental problems of the developing countries *plus* the environmental problems of the rich, industrialized ones. Some Brazilians call their country "Belindia" because of these two polar life styles, the rich living in their cozy walled enclaves as though they were inhabitants of Belgium, the underclasses living in the desperate poverty of the Indians of the Amazon basin and the dreadful shack-cities called *favelas*.

The city of Rio de Janeiro itself is a sad story of environmental decay. Even twenty years ago, the waters of its beautiful Guanabara Bay were sparkling clean for bathing, and tourists from its luxurious waterfront hotels lounged happily on its lovely Copacabana Beach. It is not the same in the '90s. Now nearly five hundred tons of raw sewage pour into the bay every day, along with another five hundred tons or more of assorted trash and waste. Ninety percent of the bay's fish are dead, and the waves now give off a faint odor of decay. During the weeks of the conference it was again possible for visitors to go to the beaches, because the authorities had chased the snatch-and-grab kid gangs that terrorized the beach areas out of town for the duration, but few wanted to. And if they raised their eyes toward the hills the visitors could see Army tanks deployed to seal off some of the *favelas*, so the city's legion of purse-snatchers and beggars could be kept temporarily out of sight.

Probably it will all get worse before it gets better. The shacks of the poor, in those *favelas* above the city, have little sewage disposal except for the open ditches at their doors, and there is no money in Brazil's cash-starved economy to pay for new sewerage and treatment facilities. Even if somehow the money could be found, says Brazilian environmentalist Roberto D'Avia, "Realistically, we are probably never again going to have dolphins playing."

So, after all that turmoil and trouble and near-billion-dollar expense, was it worth it? Did the Rio Conference achieve its purposes?

Sadly, it did not. The best that Maurice Strong, the event's secretary-general, could claim was that it had achieved "an agreement, but without commitment," while Ramshi Mayur, the environmentalist from Bombay, India, who was in attendance throughout the sessions, was more specific. Mayur said flatly, "It was a failure. We came away with nothing."

If there was one identifiable cause for the failure at Rio it perhaps belongs at the door of the United States. In the administration of the "environmental president" the environment had been getting short shrift for nearly four years. The outlook improved somewhat in the spring of 1992, perhaps because John Sununu, the least environmental official of all, had been forced to resign by public outcry over his profligate use of taxpayer money for his ski vacations. President Bush, with obvious reluctance, finally agreed to attend the Rio meeting . . . but then he made sure in his public statements beforehand that his business-as-usual supporters knew he wasn't going to *agree* to anything substantial there; and that may have been a signal for all the others who resisted change to pull out their hamstringing knives.

They hamstrung the agreements very effectively. Saudi Arabia, supported by such other oil states as Iran, Kuwait and Nigeria, fought to keep recommendations for fuel-efficient cars and alternate energy sources out of the conference's recommendations. Third World countries like India, Malaysia and China refused to accept treaty controls on deforestation because they were "infringements on our sovereignty." (But, Ramshi Mayur points out, what is any treaty if not an infringement of sovereignty? That's what treaties *are*.) And when push came to shove over the biodiversity agreement, calling for all nations to preserve endangered species, George Bush refused to let America sign it at all.

Politically that was a sort of embarrassment for the United States. The nation that for half a century had been used to setting the agenda for the Free World was isolated, as even its most loyal allies, Japan and the United Kingdom, signed the agreement—the United Kingdom with a resounding speech calling for swift and positive action on behalf of the threatened species.

But perhaps President Bush was only more candid than the others. The signing of the biodiversity agreement was more symbolic than real. Its clauses provided that it would not take any tangible effect until it had been ratified by the legislatures of at least fifty of the signatory governments, and as soon as they got home the delegates began to backtrack— the United Kingdom, for instance, deciding that it would think things over for a year before deciding whether or not to ratify, perhaps by May, 1993.

Money and Politics

When we pointed out that a lot of environmental problems were caused by bad bookkeeping—accounting that conceals the true costs of fossil fuels, for instance—we were on the right track, but now it seems that we didn't go far enough. New research by Robert Repetto, former Harvard economics professor now on the staff of the World Resources Institute, casts doubt on the reality of economic "growth" itself; an increase in Gross Domestic Product, he says, may well be a fiction that hides something close to actual bankruptcy.

In *Scientific American* (June, 1992) Repetto points out that "growth" numbers are meaningless unless they also show an expense figure for resources consumed in the process. As an example, he offers the case of a farmer who sells the timber in his woodlot and uses the money to pay for a new barn.

Of course, the farmer himself knows very well what has happened in that transaction. He is aware that he is richer for the acquisition of the barn, but also poorer for the loss of his wood. The Gross Domestic Product isn't that smart. According to its bookkeeping the sale of the wood is one bit of product, the purchase of the barn is another and, says Repetto, "Nowhere is the loss of a valuable asset reflected." What's more, if the farmer had taken the money from his lumber sale and spent it on some frivolity he would then actually be poorer, but national income would still show it as a gain.

A prime real-world example of the consequences of this fantasy-land arithmetic, Repetto points out, is the country of Costa Rica. For the sake of economic "growth" Costa Rica cleared its forests for farmland and over-exploited its rich fisheries. But now the cleared hillsides are eroding away and the fishery stocks are so depleted that an average fisherman earns less than a person on welfare, and after decades of "growth" the country is poorer than before. Much the same is true of the Philippines, Indonesia and other countries . . . and it is hard to doubt that present government policies are leading the United States down the same path.

Conservationists have always claimed that in reckless over-exploitation of our natural resources we were stealing from our children, but perhaps they were wrong. In the obsessive quest for quick profits and fictitious prosperity we are impoverishing the nation and the world; and the people we are actually embezzling from are ourselves.

We suggested earlier in this book that the quest for quick profits—especially in the junk-bond and hostile-takeover manipulations—would cause a curtailment of the very scientific research and development the world is counting on for future prosperity.

The evidence of one more year supports that notion. Japan's money woes—the deep plunge on the Tokyo Stock Exchange, the growing doubts about its unreal land prices— cost Japanese companies their own "frills," for the same reasons. Companies as large as Mitsubishi, Nissan and Fujitsu froze or actually cut their research budgets for the 1990s. Worldwide, according to *Science* in its 29 May 1992 issue, natural history museums are turning themselves into something closer to theme parks than research establishments; even in such respected institutions as the Smithsonian and both New York's and London's Natural History Museums scientific staffs have been cut, permanent positions replaced with temporaries, scientists taken off studies they have been working on for years to take up pragmatically "useful" researches. In the United States in particular, a 1992 report from the National Science Foundation shows that funding for industrial research and development rose about 7% a year between 1975 and 1985, but since has dropped to only 1.5%.

Of course, these problems are all political, and we've probably already said all we can say on the political reasons, and cures, for these misguided policies. But there is one new element that just might turn out to be a considerable help in dealing with the problems of bought legislators and welfare for the wealthy. It is a new resource, available to all of us, and it is called "Project Vote Smart."

One of the great problems in American politics is that the voters seldom know exactly who it is they're voting for. Most of the available "information" about candidates comes from TV commercials, press releases and the occasional stage-managed "debate," and most of it is an irrelevant and even pernicious as the Willie Horton ads of 1988. To find out the *truth* about a candidate takes time and effort beyond the reach of most individual voters.

In the summer of 1992 that got a lot easier, as something new came along in American politics. Sponsored by the non-partisan Center for National Independence in Politics (whose co-presidents are Jimmy Carter and Gerald Ford, and whose address is 129 NW 4th Street, Corvallis OR 97330), Project Vote Smart offers a free 800 telephone number. Any voter can pick up the telephone and quickly find out the documented facts about any candidate for federal office anywhere in the United States.

Project Vote Smart does not take sides. All it offers is information, but when you use that 800 number its batteries of volunteers and college students will dig out information for you on past voting records, biography and—perhaps most important of all—the names of corporate and individual campaign contributors. That is priceless knowledge. The private voter may not know just where a candidate stands on important issues, but the PACs and the pressure groups do, and they make their contributions accordingly.

That's the score for one more year.

Little has changed; most things have just become more so. It is still true that the basic solutions to almost all our environmental problems are at hand.

But it is also true that we have yet to show that we have the will to apply them.

Frederik Pohl
September 1992

AFTERWORD

Twenty-six Years Later

Kim Stanley Robinson

Twenty-six years later, the obvious question is: How does the book hold up? What did they get right, what wrong? What did they miss, if anything? And were their predictions accurate?

On my reading, they got almost everything right. I'm sure they wouldn't claim any great credit for this; they weren't the ones assessing the ecological situation, they were only reporting on the work of the scientific community. At the time they were writing, the situation was already clear, and only had to be articulated. As they put it at one point in the book: If you perform an amniocentesis on a pregnant woman and then declare that the child she's carrying will be a girl, this is not really an act of prediction, even if no one else is yet aware of the fact; it's just something that has already happened, discovered by indirect means. It was the same with the environmental crisis in 1991: Facts obtained by science revealed a developing situation. The only questions left were how quickly it would happen and how far it would go. Because they were writing nonfiction, Asimov and Pohl refrained from performing the science-fictional act of making any precise predictions about these particular questions; the data and modeling didn't allow it. As a result, their book is about as careful in its predictions as the Intergovernmental Panel on Climate Change's first assessment, which came out at around the same time.

It's impressive, or troubling, to see here how long we've known about most of the environmental problems we face today. Asimov and Pohl wrote of ocean acidification resulting from CO_2 uptake; reforestation as a way to capture carbon; and most importantly, economics as the key to the entire problem, because environmental damage is currently treated as an externality, thus shoving its costs onto future generations. They suggested a carbon tax as one solution to this problem, and predicted resistance to this tax (an easy call). They spoke of sea levels rising as the polar ice caps melt, but declined to guess how quickly this might happen; this is still a question beyond our ability to calculate with any accuracy. Even in 1990, they noted that the hole in the ozone layer was being mitigated by the international ban on CFCs. At the same time, they were aware that banning CFCs is a lot easier than shifting away from than fossil fuels.

So what did they miss? Nothing of importance that existed in 1991, as far as I can tell. They spent more time discussing the Gaia concept than we would now. And they gave more credit to the field of futurology than I would, calling it "a respectable, at least fairly respectable, professional discipline." Perhaps there hadn't yet been time enough to make it clear that this emerging discipline was mostly a kind of science fiction that charges customers ten thousand dollars rather than ten for their consultations. Although actually Asimov and Pohl were already pretty hip to that; they defined the first law of futurology as "the more complete and exact a prediction is, the less it is worth." In short, they knew science fiction when they saw it, but were hoping for the best when it came to futurology, because as they also wrote, "it serves a clear and present need." What I think they really meant is that futurology may be only a kind of science fiction, but science fiction itself serves a clear and present need as a way of building scenarios and modeling exercises and considering which futures we want to avoid and which to try for.

One of the most interesting sections of the book comes at the end, when they described what readers can do to help the situation. Here they covered all the bases as they saw them, including many technical, economic, and political actions that American citizens could take without completely disrupting their lives. These are for the most part acts of citizenship built into the fabric of a healthy democracy. Civil disobedience is mentioned in passing, but much more space is given to perfectly legal individual and group actions, in particular organizing into groups to leverage impacts on the political decisions that are necessary to deal effectively with environmental issues. In one respect they were again ahead of the game, and sounded fully contemporary; individual virtuous environmental actions, they wrote, are indeed virtuous, and good to do, but they won't be enough to get the job done; structural political action is required. We have to change the economic system that currently allows the planet to be degraded while profits for a small few are still declared to be somehow real. It will take an intense political effort if we are to make the fundamental economic changes we must to get the job done.

Their instructions for making a political action group, including even a blank form with its rules to be filled in as desired, speaks to the two writers' practical natures, and their liberal belief in the power of democratic action. They were neither of them quite the hothead leftists they had been when they met as teenagers, but then again they were never as radical as their friends. What they believed in, young and old, was science, democracy, and science fiction.

Their close focus on grassroots community organization is both admirable and one of the clearest signs that this book was written in an earlier time, which is to say, before globalization had taken full effect or become a well-known concept. In the immediate aftermath of the Cold War, the United States was the world's only remaining superpower, and so it made perfect sense to think that if one could change American practices, the world's environmental problems might be solved. Now, with a much more fully enacted globalization of the world's economy, which includes a greatly expanded power of unconstrained finance over all states' sovereignty, and also the rise of China, we live in a system in which climate change has become a problem that not only can't be solved by individual actions, it also can't be solved by any single country's actions, not even the United States'. It's going to take coordinated international action, which is a scary thought, because by and large we still live in a nation-state system. In that sense, the globalized economic system that has emerged might eventually be a good thing, as a kind of United Nations with teeth. If the global economic system were devoted to ecological restoration and sustainability, then everyone on Earth would be on board with the project. This has to be one of the goals of our political action now.

This new development in world history doesn't make any of Asimov and Pohl's solutions invalid, or even unimportant.

Grassroots action is still and always the real work, the work anyone can do right away. And America is still home to the majority of the world's capital and its military might, as well as a great deal of its carbon burn. So what happens here matters hugely. This twenty-seven-year-old book can be appreciated for how it speaks to that, describing what we can do here and now.

It's a book that describes an existential threat to a planetary civilization. By definition that's a job for science fiction, and these two masters of the genre do it beautifully. Do we have a dystopian problem on this planet? Yes. Is there a utopian solution? Maybe; but only if we apply science to every aspect of the problem, including the economic problem that lies at the heart of the matter. Success can only come by working with all the other members of our species to tackle the problem collectively. Can that happen? These two writers, both hard-headed realists, say yes. The teenage hope for a better future never left these guys, no matter the mounting danger they were quicker than most to see. Take the torch from them and run it forward.

Kim Stanley Robinson
January 2017

ABOUT THE AUTHORS

Isaac Asimov (1920–1992) was one of science fiction's greatest writers. Asimov's award-winning novels include *Foundation's Edge* and *The Gods Themselves*, and his work remains a profound influence on the genre.

Frederik Pohl (1919–2013) was an influential science fiction author and editor and was best known for his collaborations with other SF authors, such as Cyril M. Kornbluth on *The Space Merchants*, and Arthur C. Clarke on *The Last Theorem*.